Wetlands

Ben A. LePage

Editor

Wetlands

Integrating Multidisciplinary Concepts

 Springer

Editor
Dr. Ben A. LePage
PECO Energy Company
Environmental and Remediation Project Manager
Market Avenue 2301
S7-2, Philadelphia, PA 19103
USA
ben.lepage@exeloncorp.com

Academy of Natural Sciences
1900 Benjamin Franklin Parkway
Philadelphia, PA 19103
USA

ISBN 978-94-007-0550-0   e-ISBN 978-94-007-0551-7
DOI 10.1007/978-94-007-0551-7
Springer Dordrecht Heidelberg London New York

Library of Congress Control Number: 2011921321

Cover illustration: A photograph of a small mitigation wetland located in Hatfield Township, which is located just north of Philadelphia, illustrating the initial plantings (*Juncus effusus*) in the spring of 2009. The hydrology of this created wetland is being derived from the adjacent wetland (to the right). A photograph of the wetland a little more than one year later. Plant growth has been prolific with natural recruitment of native wetland plant species occurring in some parts of the wetland.

Cover design: deblik, Berlin

Printed on acid-free paper

Springer is part of Springer Science+Business Media (www.springer.com)

This volume is dedicated to the memory of Dr. Mark Brinson, teacher, colleague, and friend. Your work has touched us all.

Preface

I would like to begin with a quote from Keddy (2002, p. xi), "A number of recent symposium volumes on wetlands appear to be little more than expensive books with a haphazard collection of papers with no unifying theme whatsoever except for the fact that all work in wet areas". How often have I have heard this opinion iterated? The science of wetlands may indeed be best described as a loosely bound group of scientists and practitioners from diverse backgrounds that are all striving towards a common goal to preserve, maintain, and improve the world's wetlands. However, more often than not these scientists are generally working independently of one another. As such, the discipline suffers from a lack of cohesiveness and the fragmentation amongst these groups is due largely to the lack of communication and understanding between one another. This fragmentation is perhaps most apparent in the applied wetland sciences where project participants rarely work as a team on wetland projects. The result is many wetland projects end in failure. With this in mind, this volume brings together experts from diverse backgrounds in a forum where the importance of each discipline is communicated in a manner that is neither presumptive, nor so laden with jargon that the meaning of the narrative is lost to most readers. The central theme of the volume is that if we as wetland scientists want to promote and advance the science of wetlands, we need to listen to and understand what each group has to offer and begin to work as a group. Although this volume is by no means comprehensive in the disciplines that have a stake in wetland research, it does provide broad coverage across the discipline, sets the stage for collaborative work, and fills a gap in the discipline that has been identified.

The topics and scope of this book should be of interest to a wide audience including wetland scientists, geologists, ecologists, paleoecologists, policy makers, landscape architects, sociologists, engineers, lawyers, and not the least, naturalists. A brief summary of the contributions follows.

The book has been organized into three broad sections. Section 1 introduces the reader to a basic overview of the wetland sciences and some of the issues that wetland scientists face or are likely to encounter when working on wetland restoration, creation, and enhancement projects. The contributions are certainly not inclusive of all of the disciplines that would be involved in wetland projects and are intended to add new dimensions to a project at the planning stage.

Through the ages wetlands have suffered irreparable damage at the hand of man. Since European occupation we have become proficient at filling in or draining wetlands in the United States. However, over the last few decades this trend is changing, and we have made great strides towards protecting these resources. Ben A. LePage examines this transition from the old paradigm that promoted wetland eradication to the emerging paradigm that attempts to incorporate wetlands into our social fabric. In this approach science and society are working collectively to develop a multidisciplinary ecosystem-based approach for wetland restoration, enhancement and creation. One of the key elements in this approach that he identifies is the greater need for communication between project participants and the public. LePage discusses the importance of developing well defined the goals and objectives during the planning stage of wetland projects. He closes with a proposal to adopt an adaptive management strategy for wetland projects.

Marjorie L. Zeff discusses the importance of channelized flow for providing and maintaining wetland hydrology. She examines the channel morphology in a variety of wetland types to illustrate the inter-related contributions of geology, engineering, and biology to the understanding of wetlands.

Andrew deWet et al. investigate the dynamic nature of sedimentology, stream geomorphology, and the ecological integrity of fluvial ecosystems over the last several thousand years and the impacts of human activity on these processes. In their study they examined the stratigraphic record of floodplain sediment deposition and erosion surfaces of a small second order stream located in the Susquehanna River Watershed in southeastern Pennsylvania that dates back approximately 10,000 years. They document how changes in land cover and landuse since European occupation influenced the erosion and deposition of sediment in this small watershed. Their study illustrates a number of problems associated with steam restoration activities in developed and developing landscapes. They discuss how sedimentological data can be used to better understand stream responses to natural and anthropogenic changes and how to adopt and implement restoration practices that will restore ecosystem functions to small fluvial systems that are consistent with how streams functioned in the past.

Christopher J. Williams reviews the paleoecological information that is relevant to wetland restoration science. As a practioner of paleoecology he discusses some of the applied aspects of paleoecology that could be useful in a restoration sense. His focus on the use of plant remains such as leaves, seeds, fruits, wood, and pollen provides insight into the benefits and limitations of these data for reconstructing the composition and structure of ancient vegetation communities, as well as the temporal and spatial resolution of past environmental conditions. His discussion is motivated by a desire to improve the baseline conditions for restoration efforts, which in turn should improve the quality and quantity of ecosystem functions and services provided by wetland restoration.

Section 2 focuses on systems of wetland classification and the relationships between the ecological services provided by wetlands and the benefits they confer, both directly and indirectly to society. The impacts of global climate change are significant as they relate to the world's wetlands and how the manner in which the scientific community examines potential and on-going change is of considerable importance. The contributions in this section provide the reader with the information necessary to understand these issues and leave them with questions to ponder.

Wetland classification is the focus of Mark M. Brinson's contribution. He outlines the importance of how wetlands are classified from non-wetland communities and the variety of classification systems that have been developed over time to meet specific needs. The various classification systems are divided into three broad groups that are based on structural, functional, and utility criteria. The histories, benefits, and limitations of the three types of classification systems are discussed. Finally, the relationship between the utility-based classification systems and ecosystem services as well as why this type of classification system is best suited for wetland management are discussed.

John A. Nyman looks at the ecosystem services provided by wetlands, but focuses his discussion on the ecosystem services that are important to people. He expands on the utility system of classification and further refines this approach to segregate the ecosystem services that provide services that are captured by the landowner and those that are external to the landowner. His point here is that most of the economic value afforded by wetlands is not captured or external to the landowner. His presentation is centered on the importance of understanding and quantifying the ecosystem services provided by wetlands so that wetland managers and regulators can properly evaluate the effectiveness of wetland management strategies and wetland valuation methods.

Beth A. Middleton provides a thought-provoking discussion on the importance of understanding complex environmental and social relationships from a multidisciplinary point of view and in light of continued global climate change. She correctly points out that our current reductionist educational paradigm is poorly equipped to address multidisciplinary questions and in order to adapt to rapidly changing climatic and environmental conditions we as a society will need to work, think, and solve problems using multidisciplinary approaches. The discussion focuses on a few emerging programs that are truly founded on multidisciplinary platforms that are centered on global change ecology. Her discussion then shifts to examples of multidisciplinary research questions that are related to global climate change including the assisted migration of slow-moving plants to prevent extinction, long-term seed storage, and planetary scale geoengineering projects that are designed to ameliorate climate change.

Charles A. Cole and Mary E. Kentula address the sometimes problematic issues of wetland monitoring, especially the dimension of time. Most monitoring programs assume that the three wetland parameters (i.e., vegetation, soil, and hydrology) are static through time. They demonstrate that this is generally not the case and that in order to properly measure the stability of a particular parameter, the objectives of the monitoring program need to be carefully selected. They discuss the benefits and limitations of single- versus multiple-visit monitoring programs and the importance of understanding and clearly defining the objectives of the monitoring program that is to be implemented and the type and quality of the data that will be generated using a particular approach.

Human interactions play a vital role in all aspects of science, especially those like the wetland sciences that touch our lives. The loss of our wetland resources over the last few hundred years is due largely to human activities compounded by the negative perceptions associated with these ecosystems. Society dictates how we treat and manage our resources and the contributions provided in Section 3 bring to light the importance that society plays in the decision making process and the need to keep people informed and engaged in wetland projects.

Although Rachel Kaplan's contribution may at first seem far removed from the science of wetlands, her insight into the importance of communication and human behavior makes this chapter important for illustrating the need for more effective communication between the science side of the business and society. Her discussion is centered on two broad themes: information sharing and the role of nature in human well being. Information sharing is built around a framework that is called the Reasonable Person Model and it illustrates what leads people to act more reasonably when presented with information that they do not understand. Nature and human well being address the issue of perception and the importance of acknowledging and recognizing commonalities and differences in perspectives.

Morgan Robertson and Palmer Hough provide an historical overview of the practice of wetland mitigation as it relates to Section 404 of the Clean Water Act in the United States and the impacts of compensatory mitigation on wetland loss. They point out that the regulations that govern wetland loss are complicated and the notion of mitigation for temporary and permanent impacts to wetlands was not a priority for the United States Army Corps of Engineers or the United States Environmental Protection Agency until the early 1990s. More importantly, the establishment of the interim goal for the Corps of "no net loss" of wetlands in 1990 signaled a major shift in the accounting and manner that the goal of no wetland loss was to be achieved.

Royal C. Gardner and Nick Davidson look beyond our boundaries and consider wetland policy at the global level. The Ramsar convention on wetlands is an inter-governmental treaty that promotes wetland conservation worldwide. The basic tenets and duties of the signatories to this treaty are presented. In addition, they provide a detailed discussion of the obligations that parties to the convention are required to uphold. Namely, employing the "wise use" approach to wetlands; designating and conserving at least one site as a Wetland of International Importance; and international co-operation. They conclude with the benefits that have been realized in wetland conservation by Ramsar parties and the priorities that Ramsar has adopted to address emergent and future global wetland issues.

Paleontological data indicate early man has been exploiting wetlands for food or some other economic benefit for at least 8,000 years. Little has changed since then. Robert J. McInnis provides an overview on the history of wetland management with specific examples through time. His discussion highlights the close relationships between the manner in which wetlands are managed, societal values, and economic drivers. More importantly, he illustrates these values and drivers have changed through time and the resulting shifts in wetland management strategies. His discussion shifts to the twentieth century where society began to realize the importance of wetlands and the beginnings of wetland conservation strategies and programs. Today the "ecosystem approach" for wetland conservation is being adopted and promoted by various agencies worldwide. However, McInnis points out that the success of such an approach, moving forward, can only be successful through inter-disciplinary collaborations.

Nancy Minich provides a landscape architect's perspective on the qualifications, contributions, and role of landscape architects in multidisciplinary wetland projects. Landscape architects fulfill a relatively unique role in a team setting because they are trained to understand and synthesize the technical information derived from the professionals on the team (e.g., engineers, geologists, wetland sciences) and convert this information into words and concepts that are easily understood and appreciated by the public. From a public perspective, wetland projects must be aesthetically pleasing regardless of the technical and scientific merits of the project. In many cases, the landscape architect's ability to effectively interface and communicate with the public and project team has a significant impact on the success of wetland projects. Minich provides examples of a number of multidisciplinary stream resto-ration projects that have benefited from her work as a landscape architect.

Urban streams and wetlands have probably suffered the most over the last cen-tury. Limited space to re-establish the streams floodplain, the inability to stop or limit upstream pollution, as well as exorbitant cleanup costs have challenged wet-land scientists in the restoration and/or enhancement of impacted urban ecosys-tems. Lanshing Hwang and Ben A. LePage present an innovative project located along a significantly impacted tidal portion of the Anacostia River in Washington, DC where floating islands were built to create wildlife habitat and improve the aesthetics of the waterfront and water quality along this portion of the river. This project highlights what is possible using a multidisciplinary approach to address wetland issues and serves as an excellent model that can be adopted for other stream restoration projects.

Acknowledgements

Thanks are due to the following individuals Tom Hruby, Amy Jacobs, Carol LePage, Deborah Willard, and Catherine Yansa. Their peer reviews improved the quality of the manuscripts and this volume considerably.

Contents

Contributors

Mark M. Brinson Biology Department, East Carolina University, Greenville, NC 27858, USA

Charles A. Cole Department of Landscape Architecture, Penn State University, 121 Stuckeman Family Building, University Park, PA 16802, USA
e-mail: cac13@psu.edu

Nick C. Davidson Secretariat of the Ramsar Convention on Wetlands, Rue Mauverney 28, 1196 Gland, Switzerland
c-mail: davidson@ramsar.org

Andrew deWet Department of Earth and Environment, Franklin and Marshall College, Lancaster, PA 17603, USA
e-mail: andy.dewet@fandm.edu

Royal C. Gardner Stetson University College of Law, 1401 61st Street South, Gulfport, FL 33707, USA
e-mail: gardner@law.stetson.edu

Palmer Hough Office of Water, Wetlands Division, United States Environmental Protection Agency, 200 Pennsylvania Avenue, NW, Washington, DC 20460, USA
e-mail: hough.palmer@epa.gov

Lanshing Hwang Symbiosis Inc., 9008 Brae Brook Drive, Lanham, MD 20706, USA
e-mail: lanshingh@symbiosis-la.com

Rachel Kaplan School of Natural Resources and Environment, University of Michigan, Dana Building, 440 Church Street, Ann Arbor, MI 48109, USA
e-mail: rkaplan@umich.edu.

Mary E. Kentula National Health and Environmental Effects Research Laboratory—Western Ecology Division, United States Environmental Protection Agency, 200 S.W. 35th Street, Corvallis, OR 97333, USA
e-mail: kentula.mary@epa.gov

Ben A. LePage Academy of Natural Sciences, 1900 Benjamin Franklin Parkway, Philadelphia, PA 19103, USA
e-mail: ben.lepage@exeloncorp.com

PECO Energy Company, 2301 Market Street, S7-2, Philadelphia, PA 19103, USA

Erin Carlson Loy Applied Research Laboratory, Pennsylvania State University, University Park, PA 16802, USA
e-mail: emc195@psu.edu

Robert J. McInnes Bioscan (UK) Ltd, The Old Parlour, Little Baldon Farm, Little Baldon, Oxford OX44 9PU, UK
e-mail: robmcinnes@bioscanuk.com

Beth A. Middleton National Wetlands Research Center, United States Geological Survey, Lafayette, LA 70506, USA
e-mail: middletonb@usgs.gov

Nancy A. Minich NAM Planning and Design, LLC, 6575 Greenhill Road, Lumberville, PA 18933, USA
e-mail: nminich6575@comcast.net

John A. Nyman School of Renewable Natural Resources, Louisiana State University Agricultural Center, Baton Rouge, LA 70803, USA
e-mail: jnyman@lsu.edu

Morgan Robertson Department of Geography, University of Kentucky, 1457 Patterson Office Tower, Lexington, KY 40506, USA
e-mail: mmrobertson@uky.edu

Jaime Tomlinson Delaware Geological Survey, University of Delaware, 202 DGS Building, 257 Academy Street, Newark, DE 19716-7501, USA
e-mail: jamiet@udel.edu

Christopher J. Williams Department of Earth and Environment, Franklin and Marshall College, Lancaster, PA 17603, USA
e-mail: chris.williams@fandm.edu

Marjorie L. Zeff URS Corporation, 335 Commerce Drive, Fort Washington, PA 19034, USA
e-mail: marjorie_zeff@urscorp.com

About the Authors

Mark M. Brinson, Ph.D. is a Distinguished Research Professor of Biology at East Carolina University. He received his B.S. at Heidelberg College (Ohio), M.S. from the University of Michigan, and Ph.D. from the University of Florida. Research interests include the relationship of hydrologic regime to ecosystem structure and function, classification, and functional assessment of wetlands, and the effects of rising sea level on coastal wetlands. He participates in research at the Virginia Coast Reserve site of the National Science Foundation Long Term Ecological Research program and was a Fulbright Fellow in Argentina in 2002. He served as president of the Society of Wetland Scientists (SWS) and received the SWS's Merit Award in 1998. He was a member of the National Research Council (NRC) committee on Wetland Characterization, chaired the NRC committee on Riparian Zones, and serves as a member of the National Academy of Science's Water Science and Technology Board. He has provided testimony before United States Senate and House committees on the identification of wetlands. He was recently part of a team that developed an assessment of ecological condition of small coastal watersheds and their receiving estuarine waters.

Charles A. Cole, Ph.D. is an Associate Professor of Landscape Architecture and Ecology at Penn State University. He has been working with created wetlands since 1983 and with wetland hydrology since 1993. He is interested in the development processes that occur within created wetlands and how succession proceeds at all levels, from soils to vegetation. Implicit in that process is an understanding of wetland hydrology. To that end, Charles has been studying the hydrology of dozens of wetlands, many for over 10 years, developing one of the most extensive data sets on wetland hydrology in the process. In order to assess the success of created wetlands, you need to know how—and what—to monitor, hence the title of the chapter.

Nick Davidson, Ph.D. has been Deputy Secretary General of the Ramsar Convention on Wetlands since 2000, and has overall responsibility for the Convention's global development and delivery of scientific, technical and policy guidance, and advice and communications as the Convention Secretariat's senior advisor on these matters. He has over 30 years experience of research on the ecology, assessment, and conservation of coastal and inland wetlands and the ecology and ecophysiology of migratory waterbirds, with a Ph.D. (1981) from the University of Durham (UK) on this topic, and continues to publish on these issues. Prior to his current post he has worked for the UK's national conservation agencies, particularly in coastal wetland inventory, assessment, information systems and communications, and as International Science Coordinator for the global NGO Wetlands International. He is an Adjunct Professor at the Institute of Land, Water and Society, Charles Sturt University, Australia, and has been recognized by the Society of Wetland Scientists' through its International Fellow Award 2010. He has a long-standing interest in and a strong commitment to, the transfer of environmental science into policy-relevance and decision-making at national and international scales.

Andrew deWet, Ph.D. is an Associate Professor of Geosciences at Franklin and Marshall College. His research focuses on stream and groundwater processes and the use of Geographic Information Systems for the assessment of human impacts on surface processes. He has worked in Pennsylvania, Maine, and Mongolia, as well as Martian channels of disputed fluvial—volcanic origin.

Royal C. Gardner, Ph.D. is a professor of law and director of the Institute for Biodiversity Law and Policy at Stetson University College of Law. His teaching and research focus on wetland law and policy, with particular emphasis on biodiversity offsets and the Ramsar Convention. He began working on wetland policy at a national level while at the Pentagon from 1989 to 1993. In 1999–2001, Professor Gardner was appointed to the National Research Council's Committee on Mitigating Wetland Losses. He served as a member of the Ramsar Convention's Scientific and Technical Review Panel (STRP) from 2006 to 2008 and is an STRP invited expert for 2009–2012. He is the former Chair of the United States National Ramsar Committee and the recipient of the 2006 National Wetlands Award for education and research.

Palmer F. Hough, M.S. is an Environmental Scientist with the United States Environmental Protection Agency's (USEPA), Wetlands Division at the USEPA Headquarters in Washington, DC. For the past seven years, Palmer has served as the USEPA's national lead on wetland and stream mitigation issues including wetland and stream mitigation banking and in-lieu fee mitigation. Palmer was the USEPA's lead on the Nation's new *Compensatory Mitigation Regulations* that were developed jointly with the United States Army Corps of Engineers and released in 2008. Palmer has organized and conducted numerous training courses across the United States on technical and policy issues associated with wetland and stream mitigation including site selection, project planning, performance monitoring and assessment, and the long-term stewardship of mitigation sites. Before coming to Washington, Palmer worked as an Environmental Scientist in the USEPA's Region IV Office in Atlanta, Georgia, where he managed the USEPA's wetlands regulatory, enforcement, and grants responsibilities in a number of southeastern states. Palmer has a M.S. in Botany from the University of Georgia and a B.S. in Natural Resources from the University of the South.

Lanshing Hwang, M.S., L.A. is a principal of Symbiosis, Inc., scientist by training, and a passionate advocate for sustainable landscapes. Before opening Symbiosis, Inc, she worked with a range of multidisciplinary firms in the Washington/Baltimore metropolitan area and is a strong proponent of interdisciplinary collaboration. Her experience encompasses master planning, site design, and project management. Throughout her career, she has worked on projects both in the United States and abroad. Her work covers a wide spread of institutional, park, streetscape, regional planning, ecotourism, and commercial landscapes ranging from less than an acre to areas of more than 40 square miles. She is interested in bringing out the unique quality of landscapes that engages human activities by clear and succinct design.

Rachel Kaplan, Ph.D. is the Samuel T. Dana Professor of Environment and Behavior, School of Natural Resources and Environment, University of Michigan, as well as Professor of Psychology at the University of Michigan. She is also an Affiliate Professor at the College of Forest Resources, University of Washington. Dr. Kaplan is a leading authority on the importance of the natural environment in human well-being. She has developed tools that both enhance public participation in environmental decision-making and facilitate the broader understanding of environmental preference. Her diverse contributions to the field of environmental psychology appear in four frequently cited co-authored books (*Humanscape: Environments for People; Cognition and Environment: Functioning in an Uncertain World; The Experience of Nature: A Psychological Perspective;* and *With People in Mind: Design and Management of Everyday Nature*) as well as over 100 publications.

Mary E. Kentula, Ph.D. is a wetlands ecologist with the United States Environmental Protection Agency (USEPA). She is based at the USEPA's National Health and Environmental Effects Laboratory in Corvallis, Oregon. Dr. Kentula's research at the USEPA has focused on monitoring and assessment of wetlands on regional and national scales. Initially she evaluated the use of wetland creation and restoration as mitigation of wetland losses. Currently she is working approaches to assess the ecological condition of the wetland resource in support of the USEPA's 2011 National Wetland Condition Assessment. Dr. Kentula's research has been used to develop wetland policy, most recently it was cited in support of aspects of the 2008 United States federal regulations on wetland mitigation. Her work has also been recognized in the broader scientific community, including by the Society of Wetland Scientists' Merit Award in 2007 for her "outstanding contribution to wetland science" through her research on the assessment of wetlands at the watershed scale.

Ben A. LePage, Ph.D., C.S.E., P.W.S. is a Senior Remediation and Environmental Project Manager with PECO Energy Company, which serves Philadelphia and the greater area. One of his responsibilities is the management of the corporate wetland program to ensure compliance with all state and federal wetland regulations. He is an active board member on a number of watershed groups in Philadelphia and a Shade Tree Commissioner in his Township. He is a Research Associate with The Academy of Natural Sciences, an Ecological Society of America Certified Senior Ecologist, and a Professional Wetland Scientist. He has served the Society of Wetland Scientists for many years as the Chair of the Education and Outreach Committee, and recently, was elected President of the Society of Wetland Scientists for 2011–2012. He has nearly 20 years of wetlands experience in industry and academia and has worked in wetlands throughout the world. He received a B.Sc. Hon. in Biology and a Ph.D. in Geology from the University of Saskatchewan.

Erin Carlson Loy, M.S. is a Research and Development Engineer at the Applied Research Laboratory at Penn State University. She has a B.A. in Geosciences with a minor in Environmental Studies from Franklin and Marshall College and an M.S. in Civil Engineering (Water Resources) from the University of Colorado at Boulder.

Robert J McInnes, B.Sc., M.Sc., CEnv MCIWEM MIEEM is the Director of an ecological consultancy, Bioscan (UK) Ltd, where he is responsible for managing and developing their wetland workstream. Rob has over 18 years experience in the assessment, management, restoration and creation of wetland ecosystems. He has strong links with the Ramsar Convention going back several years and has represented the Society of Wetland Scientists on the Ramsar Scientific and Technical Review Panel (STRP) since 2005. On behalf of Ramsar he is leading work on urban and peri-urban wetlands where he is coordinating the production of guidelines for the integration and protection of multi-functional wetlands within urban environments. He has also reviewed and evaluated the utility and pedigree of Ramsar guidance on wetland restoration. In addition to his links with the Ramsar Convention, Rob has held several strategic roles including being President of the European Chapter of the Society of Wetland Scientists, sitting on the Technical Advisory Group of the England Wetland Vision, and being a judge on the annual Chartered Institution of Water and Environmental Management (CIWEM) Living Wetlands Award. Prior to his current position, and following a period in both academia and consultancy, Rob was Head of Wetland Conservation at a UK-based NGO, the Wildfowl & Wetlands Trust (WWT) and Managing Director of WWT's specialist consultancy.

Beth A. Middleton, Ph.D. is a research ecologist with the United States Geological Survey, National Wetlands Research Center in Lafayette, Louisiana. She has degrees in botany from University of Wisconsin (Madison; B.S.), University of Minnesota (Duluth; M.S.), and Iowa State University (Ames; Ph.D.). She served as the Society of Wetland Scientsts Secretary-General (2005–2008), and was the founder of the Global Change Ecology Section (2009). Her research is on the effects of climate change on world wetlands, and has contributed to the understanding of world wetland restoration and global climate change. Her work is well recognized and her book *Wetland Restoration, Flood Pulsing and Disturbance Dynamics* received the Merit Award of the Society of Wetland Scientists (2004). During her time as a full professor at Southern Illinois University, she held a Fulbright Fellowship at G.P Pant University, Pantnagar India on wetlands of the Himalayan terai (1991). She gave the Earth Day talk (2009) sponsored by the United States Consulate in Chennai, India. She is a visiting professor of the Chinese Academy of Science (2010) working on wetland restoration on the Sanjiang Plain of China.

Nancy A. Minich, HTR, RLA, ASLA is the principal of NAM Planning & Design, LLC, a Bucks County, Pennsylvania-based landscape architecture firm, which was formed in 2002. Nancy has a Masters degree in Landscape Architecture from the University of Pennsylvania and over 20 years of sustainable landscape architecture and planning experience. She is both a registered Landscape Architect and Horticultural Therapist. Many of her clients' projects have received awards from an array of regional environmental and planning entities and have been featured in numerous national publications and journals. She has been an adjunct professor at Delaware Valley College, Doylestown, Pennsylvania, since 1999 where she has been teaching "Ecological Landscape Management and Restoration". In addition, Nancy has been an adjunct professor in the Landscape Architecture Program at Philadelphia University since 2006 and an instructor in the Seattle–based Sustainable Building Advisory course. Nancy frequently presents at conferences and to other audiences throughout the United States on her firm's work and research in sustainability, "artful" storm water management design, and the therapeutic and social values of nature. Nancy is the Pennsylvania American Society of Landscape Architect's appointee to the National Park's Services' Historic American Landscapes Survey (HALS).

John A. Nyman, Ph.D. After earning a B.S. with a major in Biology from the University of New Orleans in 1984, Andy worked 8-days-on and 6-days-off collecting environmental data and enforcing wildlife laws at the very end of the Mississippi River until 1987. At Louisiana State University, he earned an M.S. with a major in Wildlife in 1989 and a Ph.D. in with a major in Oceanography and Coastal Sciences in 1993. After 7 years at the University of Louisiana at Lafayette, he returned to LSU in 2001 and currently is an Associate Professor in the School of Renewable Natural Resources. He has authored/co authored 38 peer reviewed publications addressing wetland loss, soils, vegetation, wildlife, fish, hydrology, and nutrient dynamics.

Morgan Robertson, Ph.D. is an Assistant Professor of Geography at the University of Kentucky. He received his Ph.D. from the University of Wisconsin—Madison in 2004, conducting a study of market and social dynamics in the creation of entrepreneurial wetland credit banking in Chicago and Minnesota. Prior to joining the faculty at Kentucky, Morgan worked in the Wetlands Division at the headquarters of the US Environmental Protection Agency, working with his co-author Palmer Hough in formulating the 2008 rule, which regulates wetland compensation under the Clean Water Act. He is the author of numerous papers on wetland banking and ecosystem service markets.

Jamie L. Tomlinson, M.S. is a Research Associate with the Delaware Geological Survey (DGS). Ms. Tomlinson's current research is focused on mapping the surface geology of Southern Delaware. Previous work at the DGS involved characterizing the subsurface geology of Sussex and Kent Counties, Delaware as it relates to the occurrence of groundwater. Her research interests outside of her primary areas of investigation include bay-bottom environments of Delaware Bay as they correlate to habitat formation and the geomorphology of modern and ancient fluvial systems.

Ms. Tomlinson received her B.A. in Geology from Franklin and Marshall College and her M.S. in Geology from the University of Delaware.

Christopher J. Williams, Ph.D. is an Assistant Professor of Environmental Science at Franklin and Marshall College. His research involves the interaction of wetland forests and climate through time. He studies modern and fossil lowland forests using the principles and techniques of neo- and paleoecology. His research interests include controls on wetland carbon cycling, paleoecosystem function, and structure of ancient wetlands, and studies of the Asian redwoods. Dr. Williams received his B.S. and M.S. degrees in Natural Resources from Cornell University and his Ph.D. in Earth and Environmental Science from the University of Pennsylvania.

Marjorie L. Zeff, Ph.D., PG., P.W.S. is a licensed Professional Geologist and certified Professional Wetland Scientist with over 30 years experience integrating her specialties of modern process geology (sedimentology and fluvial geomorphology) with wetland science (functions and mitigation site selection and design). Dr. Zeff received degrees in geology from the University of Rochester (BA), Duke University (MS), and Rutgers University (PhD). Her doctoral research findings on salt marsh tidal channels have been published in *Marine Geology* and *Restoration Ecology*. Dr. Zeff is currently employed with URS Corporation in Fort Washington, Pennsylvania.

Part I
Introduction

Chapter 1
Wetlands: A Multidisciplinary Perspective

Ben A. LePage

Abstract Today wetland managers and practitioners are faced with considerable challenges and responsibilities. Not only do they need to consider and fully understand all of the traditional components of wetland science such as ecology, geology, and engineering, but also, the aesthetic, socio-political, and economic aspects of the discipline. Nevertheless, over the last several decades the science of wetlands in the United States has matured and significant progress has been made in the area of wetland creation, restoration, and enhancement. However, despite changing attitude towards wetlands many projects still end in failure; a result due largely to adherence to outdated process-driven models, the lack or inability to effectively communicate to project team members and society, and the continued compartmentalization of the various disciplines and practitioners involved in wetland projects. Here the contributions and opportunities that the diverse disciplines bring to the science are illustrated and the importance of effective communication within a proposed adaptive management strategy is presented and discussed. This new adaptive management strategy will allow wetland practitioners to move away from prescribed process driven approaches such as our "no net loss" policy, to one that is more flexible and adaptive to society's needs. As the emerging greenhouse gas market develops new wetland valuation methods will emerge and wetland practitioners, society, and the regulatory agencies will need to work collaboratively to react quickly and strike a balance between creating wetlands that maximize profitability with respect to the carbon and nitrogen markets and wetlands that provide other ecosystem services.

It is estimated that man has successfully eliminated more than one half of the world's wetlands and a large percentage that remain are significantly impacted by human

B. A. LePage (✉)
Academy of Natural Sciences, 1900 Benjamin Franklin Parkway, Philadelphia, PA 19103, USA
e-mail: ben.lepage@exeloncorp.com

PECO Energy Company, 2301 Market Street, S7-2, Philadelphia, PA 19103, USA

B. A. LePage (ed.), *Wetlands,*
DOI 10.1007/978-94-007-0551-7_1, © Springer Science+Business Media B.V. 2011

activities. In the United States, urban and agricultural development reinforced by negative public perceptions have contributed to the loss of approximately 55%, or 122 million acres, of the estimated 221 million acres of wetlands that existed prior to arrival of Europeans (Dahl and Allord 1996). Only now are we beginning to realize that our survival and way of life is inextricably linked to the wetlands that we sought and continue to destroy. This trend is reversing however, and for the first time in 300 years, at least in the United States, we are creating, restoring, and enhancing more acres of wetlands than we are destroying (Dahl 2006). Between 2004 and 2008 federal agencies reported that 1,197,000 acres of wetlands were created and restored. Of this amount, approximately 5% or 59,850 acres, (14,962 acres per year), were created (Council on Environmental Quality 2008). This amount of created wetlands should however be considered a minimum because it does not include the wetlands that are being created in response to State regulatory programs and therefore underestimates the total amount of wetlands created in the nation. While this trend in the United States is promising, habitat destruction due to infrastructure growth, especially in developing countries, is placing additional pressure on biodiversity and habitat loss (World Bank 2007; Kiesecker et al. 2010).

In the last 30 years the science of wetlands has matured and in recent years our understanding of these complex ecosystems has improved considerably. The movement to protect our wetland resources, as well as stricter adherence to Federal and State wetland regulations, has contributed to a notable increase in wetland creation, enhancement, and restoration projects throughout the country especially in urbanized areas. The notion that the study of wetlands requires "a multi-disciplinary approach or training in a number of fields not routinely combined" (Mitsch and Gosselink 1986, p. 11) remains one of the cornerstones of the discipline, but one that has not been truly embraced. Today's wetland creation, enhancement, and restoration projects are complex, expensive, and require the seamless integration of ideas and contributions from a team of individuals with expertise from diverse and seemingly unrelated disciplines if they are to succeed.

While there is considerable literature available on different facets of wetlands, much of it is technical and targeted for specific audiences (e.g., France 2002; Connolly 2006; Dunnett and Clayden 2007; Kadlec and Wallace 2008; Vymazal and Kröpfelová 2008). For example, ecologists learn about wetland ecology, engineers about treatment wetland engineering, and so on, but no where do ecologists have the opportunity to learn about engineering or vice versa. The content of these publications is arguably difficult to understand, or at least challenging for non-specialists, and project managers generally fail to fully appreciate the value of insight and contributions that other disciplines could bring to a project. The result is that the traditional boundaries and philosophies between the participating disciplines that contribute to the diverse wetland practices are still very much compartmentalized, which in turn impedes the progress and success of such projects.

Perhaps the greatest obstacle to removing the barriers that contribute to compartmentalization is the lack of or ineffective communication between disciplines. As pointed out by Kaplan (2011), communication needs to overcome a number of barriers to elicit a level of understanding or consensus among all participants involved.

Though the barriers between disciplines are beginning to be dismantled, there are still many instances where project participants work in a vacuum, rarely interacting with other team members.

With this point in mind, the purpose of this volume is one that attempts to illustrate the multidisciplinary nature of the wetland sciences to a broader audience of current and potential wetland practitioners. It is also meant to highlight the contributions and opportunities that these diverse disciplines bring to the science. While this volume is in no way meant to be comprehensive with respect to the various disciplines that are highlighted here, it does demonstrate that wetland science contributors are diverse and encompass a broad range of scientific and social disciplines. Today wetland managers are being faced with considerable challenges and responsibilities. Not only do they need to consider and fully understand all of the traditional components of wetland science such as ecology, geology, and engineering, but also, the aesthetic, socio-political, and economic aspects of the science. These factors all contribute and play a key role in the success of a project. To develop a multidisciplinary, ecosystem-based strategy for success, project teams must embrace the rich diversity and valuable contributions of the participants and stakeholders that are listening, communicating, and understanding one another. The social aspect, which has previously not been considered to be important, is perhaps now the major stakeholder in decision making processes (see McInnis 2011). So project teams not only need to re-define themselves in terms of their structure, how information is being processed, and how decisions are made, each step of a project now must integrate the needs, responses, and contributions of society.

The contributors to this volume have written about their areas of expertise in a manner that will provide wetland team participants, aspiring wetland scientists, and non-professionals, with cogent discussions of the importance of their disciplines that extend beyond the immediate or traditional perceptions of those disciplines. Readers will note that some of the contributions that comprise this volume are targeted at an applied audience. This strategy is deliberate for two reasons. First, we are really at a stage in the evolution of the discipline where the emphasis is moving from a more prescribed or process driven approach, to one that is much more flexible and adaptive to society's changing conditions and needs. Second, if we are to maintain the no net wetland loss policy, at least in the United States, we will eventually reach a point where there will be fewer wetland enhancement and restoration projects and the only opportunity for stemming wetland loss will be projects involving created wetlands. But before we reach this point, we need to understand what types of wetlands are being created and why they are being created. As governments move towards a more globalized approach and markets for carbon and nitrogen sequestration become established, a balance between the types of wetlands created must be struck, otherwise we are likely to experience ecological and economic issues similar to those seen with agricultural and forestry monocultures. This approach however, does not diminish the importance of wetland enhancement and restoration projects and an adaptive multidisciplinary management strategy is strongly recommended for all wetland projects.

Wetland Purpose and Planning

The types of wetlands recognized vary with the system of classification one adopts, but all classification systems encompass a wide range of ecosystems (biological communities), forming a continuum from small, simple, and ephemeral to large, complex, and geographically extensive communities. Regardless of their size, all wetlands serve a purpose, but in many cases this purpose may not be fully realized because not all wetlands perform the same function(s), and if they do, it is unlikely that these functions would be performed at the same level. More importantly though, if the function(s) provided by wetlands are not recognized by society as being important or fail to impart some economic benefit to a landowner, the wetland is considered worthless. Over the last several decades the scientific community, as well as a multitude of special interest groups have identified a great number of ecological and societal benefits that wetlands provide. These benefits, commonly referred to as ecosystem services, are the direct or indirect contributions that ecosystems make to the well-being of human populations (United States Environmental Protection Agency 2009). Some are quantifiable, while others such as aesthetics are not. Regardless of whether they are based in science or simply allow one to recall happy childhood memories, they are all important and no one aspect should be considered more important than another. While philosophically this may be the case, in reality, services that can be bought and sold are valued higher than those functions that are external to the landowner and have no perceived or market value. The concept of market value includes the goods, services, and assets that are traded in markets and ranges from traditional goods and services such as timber to emerging commodities such as greenhouse gas offsets and nitrogen mitigation (Murray et al. 2009). What is important to remember is that the value of the commodities that are being bought and sold is tied to societal needs and is likely to vary across regions and change over time (McInnis 2011; Nyman 2011). Over the last few decades society has become much more environmentally conscious and we are now becoming much more aware of our environmental surroundings, the impacts of human activities on ecosystems, and the long-term consequences of our activities (e.g., Millennium Ecosystem Assessment 2005). History has shown that small changes in climate can significantly impact human civilization by eliminating wetlands (Middleton 2011). Is history repeating itself?

Today, a reasonable barometer of society's attitude towards its ecological resources is the health of these resources. Although we are now beginning to recognize the ecological value of wetlands and taking steps to implement more stringent measures to preserve and improve the quality of the wetland resources that remain, the shift in society's attitude towards wetlands is still very much in transition. Continued ecological impacts and new challenges such as global warming, rising sea levels, major oil spills, and disease (e.g., West Nile virus, malaria) only add to the temporal incongruity between knowing what we need to do and doing what we need to do. No matter how difficult the challenge, wetland scientists must remain committed to advancing the science and educating the public. This may be the right time to develop new approaches to deliver information on wetlands to the public if we are to effect changes in society's behaviors and attitudes more quickly.

However, as alluded to previously, before we embark on a mass campaign for wetland creation, restoration, and enhancement for the sake of meeting the goal of no net wetland loss, the notion of purpose must be considered. Wetlands provide a number of important functions such as sediment/toxicant retention, nutrient removal and/or transformation, groundwater recharge/discharge, floodflow alteration, wildlife habitat, and recreation to name a few (Keddy 2000; Nyman 2011). Nahlik et al. (2010) provide a summary of the ecosystem services that have so far been described in the literature. They categorized twenty-five functions into five broad groupings based on the services that they provide. These groupings include structure, functions, processes, goods, and uses and each group includes a list of specific ecosystem services and each group is further sub-divided into specific ecosystem services (Table 1.1).

The functions provided by wetlands depend on factors such as age, size, and structural complexity of the wetland (Mitsch and Gosselink 1986). It is important to point out here that the concept of wetland health is somewhat misleading. In order to properly assess such measureable functions, they must be compared to a similar

Table 1.1 Ecosystem services. (Modified from Nahlik et al. 2010)	Major service types	Specific ecosystem services
	Structure	Biodiversity
		Wildlife habitat
	Functions	Carbon sequestration
		Water purification
		Nutrient retention
		Flood prevention
		Fisheries support
		Food provisioning
		Soil development
		Storm mitigation
		Rare species support
		Waste assimilation
		Climate change mitigation
	Processes	Nutrient cycling
		Primary productivity
		Nitrogen removal
	Goods	Grains, tuber, fibers
		Fish, shellfish, crustaceans
	Uses	Recreation
		Spiritual practices
		Human well-being
		Bird watching
		Inspiration
		Ecotourism
		Relaxation

wetland that is less impacted. While this approach, in principle seems valid, there are few remaining wetlands that have not been impacted by human activities and continued use of this approach ultimately diminishes the quality of the wetland services being measured. It is analogous to making a photocopy of a photocopy and so on. At some point the photocopy no longer resembles the original. The question of wetland health should therefore be more appropriately addressed by asking "What is the potential of the wetland to perform a particular function?"

Therefore, with respect to wetland creation, restoration, and enhancement we need to be clear on the short- and long-term objectives that we are attempting to achieve. We need to be clear on whether we are creating, restoring, and enhancing wetlands to achieve specific ecological functions or meet acreage quotas? Economics will also play a major role. The emergence of nutrient and carbon markets holds great promise for wetland projects, but at what cost given that such markets are typically driven by maximizing profits at the expense of creating a quality or sustainable product. Are we creating more problems for ourselves by restoring wetlands that are recovering on their own? There are many cases where previously impacted wetlands are recovering and the contaminants that once negatively impacted the biota (plants and animals) are no longer available to plants and animals. This process is called natural attenuation. Should we be focused on restoring these types of wetlands or just let them continue to recover on their own? There are clearly more questions than answers and the regulatory agencies together with stakeholders need to consider scenarios such as these if the ultimate goal is to create profitable and healthy self-sustaining wetlands.

Today's wetland projects typically require a significant investment of time and resources. Planning is of the utmost importance. There is no point of creating, restoring, or enhancing any wetland if it cannot become sustainable, while providing the maximum number of ecological services for which it was designed or at least capable of providing. What about the aspect of time? While the potential ecological services afforded by a specific type of wetland needs to be initially integrated into the local and regional landscape, wetland communities are dynamic. They develop and evolve in response to climate and environmental change and their character will change over time. Do we plan for a static endpoint in the design with maintenance and monitoring occurring in perpetuity or allow the wetlands to reach a dynamic equilibrium with their surroundings, regardless of the endpoint?

Wetland restoration, creation, or enhancement requires considerable "up-front" planning to get to a conceptual stage where a project team has an agreed strategy and direction. However, integrating the conceptual design into the physical setting is a step that is often not part of the planning process. This is perhaps one of the main reasons that many wetland projects end in failure. Understanding of the local and regional landscape as well as the hydrology and geology provides limitations on the types of wetlands that can be supported in specific physical settings as well as their potential to provide certain ecosystem services. Site geology is a factor that is as important as hydrology. The local geology impacts the hydrologic regime (surface and groundwater flow), soil composition and chemistry (minerals, nutrients, and pH), and to some extent the structure and composition of the vegetation (see Zeff 2011). From the standpoint of a scientifically sound wetland management strategy, the front end investment in time and resources that allows project members to fully understand the physical attributes of a site during the pre-design phase is priceless. Knowing and understanding the types of wetlands that can be supported at a particular location is a fiscally and ecologically responsible strategy that offers wetland managers the best opportunity for successful wetland creation, enhancement, and restoration projects.

Without suitable hydrology, wetlands simply would not exist. Superimpose hydrology on diverse landscapes and the result is a series of distinct wetland types (Brinson 1993). Change the landscape or hydrologic regime and the character of the wetland changes in response to the new conditions. For created wetlands the success or failure of a wetland depends entirely on how well the hydrology and geomorphology (the physical setting) have been engineered and integrated into the overall site design. Obtaining a clear picture of the hydrogeomorphic setting of a project site is one of the most crucial steps in the planning phase of a project. Nevertheless, as is the case with many projects, most do not go as planned. Emergent issues that have been identified through wetland monitoring studies of constructed and natural wetlands from the Pacific Northwest include atypical hydrogeomorphology and plant responses to hydrologic change (Gwin et al. 1999; Magee et al. 1999; Magee and Kentula 2005). The results of these studies illustrate the need for careful and detailed planning if wetland managers are to develop and maintain self-sustaining native wetland communities.

The idea of whether a project is "successful" brings up two important points. As Brinson (2011) points out, humans have an innate need to classify things based on a set of criteria. In most cases wetland classification schemes generally fall into three buckets based on their structure, their functioning, and their utility. While there may be similarities between various classification schemes, each serves a specific purpose. As Gwin et al. (1999), Magee et al. (1999), and Magee and Kentula (2005) pointed out, the constructed wetlands that they studied did not fit into one of the pre-defined hydrogeomorphic categories. This however, should not diminish the value of these wetlands. The point here is that enhanced, restored, and especially constructed wetlands may not necessarily fall into a pre-defined box that we use to classify naturally-occurring wetlands. We should not be focused on trying to deem the success of a project based on whether it fits in a pre-defined box, but rather we should be focused on measuring the potential value in terms of ecosystem services that these wetlands provide.

The second point that has bearing on the success of a project, especially those mandated by State and Federal agencies, is the notion of measurement. For a project to be considered successful certain aspects of a wetland need to measured and compared against a standard. Typically, a less impacted reference wetland that is located near the mitigation wetland is used as a benchmark. Maintenance and monitoring requirements usually extend for a five-year period. If the mitigation wetland achieves and maintains certain metrics compared to the reference wetland over a specified period of time, then the project is deemed to be a success. However, Cole and Kentula (2011) indicate that the metrics used by most regulatory agencies for maintenance and monitoring programs typically measure community structure and not wetland function. Perhaps it is time for all stakeholders involved in wetland mitigation to come together and develop a new integrated wetland management strategy that relies less on structure and adopt an approach that is founded on measurement of their potential for ecological services.

Wetland Management

Most wetland management strategies follow a prescriptive approach to maximizing or protecting a single ecological service without considering the impacts to these ecosystems as a whole (Euliss et al. 2008). In many cases the implementation of procedures and policies aimed at protecting and enhancing one or a few ecological services has had a negative impact on the remaining services. As a result, it is unlikely that we will ever come close to restoring the cumulative ecological functions in the United States that were provided by the pre-European wetlands. While our goal of no net wetland loss in the United States is noble, we do appear to be sacrificing wetland quality for quantity.

On one hand, society has developed ecological amnesia because of its shortsighted approach to restoration. Our inability to remember what a pristine wetland is supposed to look like has led to the practice of establishing diminishing ecological restoration benchmarks. This is most prevalent in more urban areas where we have replaced our higher-quality forested and scrub-shrub wetlands with palustrine emergent wetlands and open water (Fig. 1.1). Unless we consciously develop and strive for restoration benchmarks and develop better long-term strategies that provide considerably higher ecosystem services than those in use today, it is not likely that we will be able to maintain the ecological quality of wetlands that remain. On the other hand increased globalization and horticultural practices have contributed to the spread of exotic and invasive plant species at an unprecedented rate and we may be at risk of homogenizing our wetland and terrestrial communities. While the spread of invasive species (both plant and animal) is perhaps one of the most pressing ecological issues, economic policy overrides enforcement and the problem of invasive species remains unabated.

One of the goals of wetland projects is to increase biodiversity while maintaining healthy local populations of native species. Is this a reasonable goal? Certainly, we are capable of creating incredibly diverse wetlands, but at what cost? Should we set unreasonable expectations for wetland projects to maintain the sometimes unrealistic vision of our policy makers and stakeholders? Sometimes a *Typha* (cattail) wetland is just a *Typha* wetland. While the importance of striving for increased wetland biodiversity should not be diminished, these questions are fundamentally important and need to be reconciled at the planning stage of a project once the type of wetland that can be supported by the regional landscape has been identified.

Wetlands are dynamic ecosystems and respond to the prevailing environmental conditions and physical setting and are susceptible to climatic change. Therefore,

Diminished Restoration Benchmarks

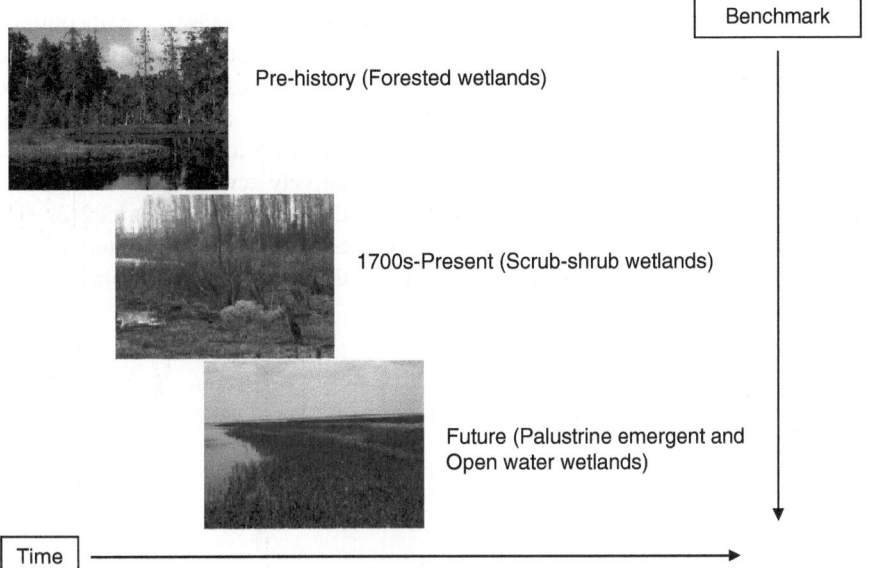

Pre-history (Forested wetlands)

Benchmark

1700s-Present (Scrub-shrub wetlands)

Future (Palustrine emergent and Open water wetlands)

Time

Fig. 1.1 Image illustrating the concept of diminishing ecological restoration benchmarks where forested and scrub–shrub wetlands become replaced with palustrine emergent wetlands and ultimately open water wetlands. Though the acreage may remain consistent through time, the ecological services provided by the palustrine emergent and open water wetlands are significantly lower than those provided by the forested and scrub–shrub wetlands

effective management strategies should be based on a dynamic model that considers both short- and long-term biotic (biological) and abiotic (physical) changes. To accomplish this, wetland managers are urged to consider the expertise of paleontologists and geologists during the pre-design phase. The data preserved in ancient wetland soils often provide spectacular spatial and temporal records of shifting patterns of vegetation change in response to climate and environmental change. While this approach is not meant to create or return wetlands to the regions pre-historic conditions, these data provide wetland managers with insight on the probability of success for a chosen end product given the existing physical and climatic parameters. In addition it provides a snapshot of the potential for the short- and long-term stability of the final wetland design (LePage 2007a, b; de Wet et al. 2011; Williams 2011). It would be unrealistic to create a wetland that is based entirely on the fossil record and expect the structure and species composition to remain static in a continuously changing environment without expending considerable time and resources in the management of the wetland. This approach is not sustainable and a poor use of limited financial resources. de Wet et al. (2011) and Williams (2011) provide specific examples that utilize paleontological and sedimentological approaches to address issues such as wetland stability over time. They also illustrate the value and limitations of these types of data and approaches for wetland creation and restoration projects during the planning stages.

As the world's population continues to grow, the urban landscape will continue to impinge on natural areas, including wetlands. Human activities typically have a direct and immediate negative impact on our natural resources. In addition to the suite of ecological processes that must be managed, aspects such as recreation, socio-political, and economic factors need to be included as part of the management strategy. While such a concept is relatively new to the discipline, this approach has been successfully developed in other areas such as urban forestry (Bradley 1995) and we need look to the successes and failures encountered in the development of these management strategies in the process of developing one for ourselves.

The Social Variables

Economics

Market values on society's goods and services form the foundation of wetland valuation. In the United States, the value of wetlands is primarily based on the market value of land and its capacity for development … a commodity that we currently consider to be important enough that a market for developable land exists. Although our understanding of the benefits that wetlands provide as a whole is shifting, the perception that wetlands are worthless has not yet been completely eradicated from society's attitude on wetlands. As a result, the direct market value for the ecological services provided by wetlands (Table 1.1) that are external to landowners such as clean water, carbon sequestration, nitrogen mitigation, sediment retention, flood control, or biodiversity have been ignored. Markets associated with carbon sequestration and nitrogen mitigation are however beginning to emerge. While the cost for clean water and flood control are measurable and could be incorporated into existing valuation models, we have yet to address how to place a market value, as opposed to an economic value, on abstract concepts such as ecological biodiversity or aesthetics.

Generally ecological services are not included in wetland valuation models because there is either no perceived value for a particular function or there is no market for the services provided by that function. Nyman (2011) focuses on ecological values that are captured by the landowner as opposed to those that are external to the landowner. He indicates that most wetland owners currently do not benefit from the economic benefits provided by their wetlands. However, the emerging market demand for carbon credits may begin to provide economic value to wetland landowners and managers for ecological functions such as wildlife habitat, water filtration, and nutrient recycling that in the past were not considered to be economically important. As is now stands our ability to truly integrate the value of most eco-system services and develop a wetland valuation model that is more ecologically grounded, rather than the current commodity-based land development approach, is limited. We may however be on the cusp of significant change. The importance of including all ecosystem services in a wetland valuation system is being recognized and efforts are being made to better define and quantify these services (Barbier et al. 1997; Boyd and Banzhaf 2007; Fisher et al. 2009; Pendelton 2009; United States Environmental Protection Agency 2010).

Institutional

Legal, political, administrative, and economic issues play an important role in shap-ing how information on wetlands is presented to the public. Use of these channels must convey the correct information to the public, while still providing the flex-ibility to incorporate, modify, and communicate new and possibly contradictory findings. As Bradley (1995, p. 9) points out "… it is important to know how people learn, why they show interest in some, but not other information, and the process that individuals go through before they find something to be important". Generally change occurs only when people believe their lives are being impacted positively or negatively. Acceptance of new information will be expedited by demonstrating direct negative or positive impacts to people's lives. The wetland community will need to work closely with regulatory agencies and interested stakeholders to de-velop educational programs and capitalize on partnerships to present wetland infor-mation for public consumption.

Unfortunately in the United States, Section 404 of the Clean Water Act (CWA), which regulates and protects wetlands, is exceptionally complex, open to interpre-tation, and continues to allow wetland loss through the process of mitigation. As pointed out previously in this chapter, wetland replacement (on a per acre basis) through mitigation does not always equate to an equivalent replacement of ecosys-tem services. Yet, this approach continues to be the *modus operandi* of our Federal and State regulatory agencies. Recent regulatory changes at the Federal and State levels, however, indicate the current administration is taking environmental issues more seriously than in the past and attempting to plug the regulatory holes. A more detailed discussion of the issues associated with the CWA is presented in Robertson and Hough (2011).

On a global scale the Ramsar Convention on Wetlands is an intergovernmental treaty that promotes international cooperation aimed at the conservation and wise use of wetlands and their resources (Gardner and Davidson 2011; Ramsar 2010). Its mission is to "… develop and maintain an international network of wetlands which are important for the conservation of global biological diversity and for sustaining human life through the maintenance of their ecosystem components, processes and benefits/services" (Ramsar 2010). To date, there are 160 contracting countries and

1,913 wetlands of international importance with a surface total of 186,982,227 ha. The United States became a party to the convention in 1987 and currently has designated 29 wetlands for inclusion in the *List of Wetlands of International Importance* with a surface area of slightly over 1.4 million ha. Although this figure appears large, it only represents a disappointing 0.77% of the total surface area of all designated sites and 28th among the participating 160 nations (Table 1.2).

Table 1.2 Ramsar participants

Country	Number of sites	Hectares	Percent of total surface area
Canada	37	13,066,675	7.02
Chad	5	12,405,068	6.66
Russian Federation	35	10,323,767	5.54
Congo	7	8,454,259	4.54
Sudan	4	8,189,600	4.40
Mexico	113	8,161,357	4.38
Bolivia	8	7,894,472	4.24
Australia	65	7,510,177	4.03
Democratic Republic of Congo	3	7,435,624	3.99
Peru	13	6,784,042	3.64
Brazil	11	6,568,359	3.53
Guinea	16	6,422,361	3.45
Botswana	1	5,537,400	2.97
Argentina	19	5,318,376	2.86
United Republic of Tanzania	4	4,868,424	2.61
Niger	12	4,317,589	2.32
Mali	1	4,119,500	2.21
Zambia	8	4,030,500	2.16
France	36	3,314,275	1.78
China	37	3,168,535	1.70
Algeria	47	2,981,160	1.60
Gabon	9	2,818,469	1.51
Denmark	38	2,078,823	1.12
Kazakhstan	7	1,626,768	0.87
Iran	22	1,483,824	0.80
United States	26	1,442,618	0.77
Mongolia	11	1,439,530	0.77
Pakistan	19	1,343,627	0.72
United Kingdom	168	1,274,323	0.68
Mauritania	4	1,240,600	0.67
Togo	4	1,210,400	0.65
Cuba	6	1,188,411	0.64
Benin	4	1,179,354	0.63
Nigeria	11	1,076,728	0.58
Former USSR	4	929,700	0.50
Germany	34	868,226	0.47
Netherlands	49	818,908	0.44
Finland	49	799,518	0.43

Table 1.2 (continued)

Country	Number of sites	Hectares	Percent of total surface area
Madagascar	6	787,555	0.42
Paraguay	6	785,970	0.42
Cameroon	5	784,115	0.42
Ukraine	33	744,651	0.40
Tunisia	20	726,541	0.39
Mozambique	1	688,000	0.37
Romania	5	683,628	0.37
India	25	677,131	0.36
Indonesia	3	656,510	0.35
Burkina Faso	15	652,502	0.35
Kyrgyz Republic	2	639,700	0.34
Namibia	4	629,600	0.34
Guatamala	7	628,592	0.34
Bangladesh	2	611,200	0.33
Papua New Guinea	2	594,924	0.32
Uzbekistan	2	558,400	0.30
South Africa	20	553,178	0.30
Sweden	51	514,675	0.28
Costa Rica	11	510,050	0.27
Armenia	2	492,239	0.26
Columbia	5	458,525	0.25
Uganda	12	454,303	0.24
Uruguay	2	424,904	0.23
Nicaragua	8	405,691	0.22
Central African Republic	2	376,300	0.20
Thailand	10	370,600	0.20
Sierra Leone	1	295,000	0.16
Belarus	8	285,807	0.15
Spain	63	281,768	0.15
Morocco	24	272,010	0.15
Turkmenistan	1	267,124	0.14
Venezuela	5	263,636	0.14
Hungary	28	235,430	0.13
Estonia	13	225,960	0.12
Malawi	1	224,800	0.12
Honduras	6	223,320	0.12
Ecuador	13	201,126	0.11
Panama	5	183,992	0.10
Turkey	13	179,898	0.10
Ghana	6	178,410	0.10
Greece	10	163,501	0.09
Chile	9	159,154	0.09
Latvia	6	148,718	0.08
Poland	13	145,075	0.08
Iraq	1	137,700	0.07
Equatorial Guinea	3	136,000	0.07

Table 1.2 (continued)

Country	Number of sites	Hectares	Percent of total surface area
Malaysia	6	134,158	0.07
El Salvador	4	133,326	0.07
Japan	37	130,027	0.07
Cote d'Ivoire	6	127,344	0.07
Austria	19	119,962	0.06
Norway	37	116,369	0.06
Eqypt	2	105,700	0.06
Kenya	5	101,849	0.05
Senegal	4	99,720	0.05
Azerbaijan	2	99,560	0.05
Liberia	5	95,879	0.05
Moldovia	3	94,705	0.05
Tajikistan	5	94,600	0.05
Portugal	28	86,581	0.05
Croatia	4	86,579	0.05
Albania	3	83,062	0.04
Marshall Islands	1	69,000	0.04
Philippines	4	68,404	0.04
Ireland	45	66,994	0.04
Italy	51	60,052	0.03
Iceland	3	58,970	0.03
Bosnia and Herzegovina	3	56,779	0.03
New Zealand	6	55,512	0.03
Czech Republic	12	54,681	0.03
Cambodia	3	54,600	0.03
Serbia	3	53,714	0.03
Lithuania	5	50,451	0.03
Seychelles	3	44,022	0.02
Belgium	9	42,938	0.02
Slovakia	14	40,697	0.02
Guinea-Bissau	1	39,098	0.02
Jamaica	3	37,765	0.02
Georgia	2	34,480	0.02
Nepal	9	34,455	0.02
Bahamas	1	32,600	0.02
Gambia	3	31,244	0.02
Vietnam	2	25,759	0.01
Belize	2	23,592	0.01
Macedonia	2	21,616	0.01
Bulgaria	10	20,306	0.01
Montenegro	1	20,000	0.01
Comoros	3	16,030	0.01
Trinidad and Tobago	3	15,919	0.01
Suriname	1	12,000	0.01
Syrian Arab Republic	1	10,000	0.01
Switzerland	11	8,676	0.00

Table 1.2 (continued)

Country	Number of sites	Hectares	Percent of total surface area
Sri Lanka	3	8,522	0.00
Republic of Korea	12	8,218	0.00
Slovenia	3	8,205	0.00
Jordan	1	7,372	0.00
Bahrain	2	6,810	0.00
Antigua and Barbuda	1	3,600	0.00
Djibouti	1	3,000	0.00
Cyprus	1	1,107	0.00
Lebanon	4	1,075	0.00
Burundi	1	1,000	0.00
United Arab emirates	1	620	0.00
Fiji	1	615	0.00
Palau	1	439	0.00
Lesotho	1	434	0.00
Mauritius	2	379	0.00
Israel	2	366	0.00
Luxenbourg	2	313	0.00
Myanmar	1	256	0.00
Liechtenstein	1	101	0.00
St. Lucia	2	85	0.00
Libyan Arab Jamahiriya	2	83	0.00
Barbados	1	33	0.00
Monaco	1	23	0.00
Sao Tome and Principe	1	23	0.00
Malta	2	16	0.00
Lao People's Democratic Republic	2	–	0.00
Cape Verde	3		0.00
Rwanda	1		0.00
Samoa	1		0.00
Yemen	1		0.00

Given that the Ramsar Convention is not regulatory in nature, are there then tangible benefits of being a party to the Convention? In the United States, benefits identified by Gardner and Davidson (2011) include continuation of the ecological character of a site by communicating the site's value to the public, influencing landuse and development activities, focused attention on long-term stewardship, increased community support, and access to financial resources to name a few.

Stormwater Management

While the role of native wetlands in the landscape is well defined, that of the urban wetlands is not because of our limited understanding of multi-purpose wetlands in a space-constrained landscape and little experience communicating the information that we have to the public. Many of our current wetland management practices are based on nearly 300 years of information and attitudes that contradict what we know to be true. Touting the benefits of wetlands and changing public perceptions is not easily accomplished and will certainly require creative approaches. For example, the Philadelphia Water Department (PWD) is focused on improving water quality in its watersheds and has recently taken innovative steps to address the issue of stormwater runoff by implementing a new cost structure to its customers for stormwater that enters its treatment system as a result of impervious cover. Customers wishing to reduce their monthly stormwater charges have the option of implementing structures or approaches (referred to as Best Management Practices or BMPs) aimed at reducing runoff ideally through infiltration. Homeowner BMPs that are easily implemented and cost effective include the installation of rain barrels, porous pavement or paver stones, vegetative buffers, and the creation of mini wetlands called rain gardens. The idea here is to put more water back into the ground through wetlands or rain gardens, rather than having it flow into the city's water treatment system. To address the more serious issues of water quality and stormwater management on a larger scale, the PWD has embarked on building multi-purpose urban wetlands. Saylor Grove is an excellent example of a small (0.70 acre) urban treatment wetland that was constructed to improve water quality by filtering point and nonpoint source pollutants, reduce streambank erosion, and desynchronize peak storm water flows and volumes into Monoshone Creek. The wetland filters approximately 70 million gallons of water that it receives from the 156 acre watershed. Completed in 2006, community residents now care for this wetland and utilize Saylor Grove for recreation and education (Fig. 1.2).

Communication

In the United States over 75% of the population resides in cities (Nowak et al. 2000). As a result, the population is becoming increasingly more disconnected from natural resources and over time the benefits that these resources provide are forgotten. Given the opportunity many people choose to leave the city to be with nature and "recharge their batteries". Natural environments have the capacity to bring out the best in people (Kaplan and Kaplan 2008), yet we continue to build developments that eradicate nature. That we spend the work week negatively impacting our natural resources and then instinctively seek them to satisfy our need to be one with nature on the weekend is a paradox.

Understanding how we think, react, and behave is central to developing an effective wetland communication and education model. Kaplan (2011) provides a comprehensive discussion on the importance of more effective and less frustrating information sharing using a framework called the Reasonable Person Model (RPM). The RPM fosters reasonableness in people by: (1) focusing on people's need to understand what is going on around them; (2) the ability for people to use

Fig. 1.2 Photographs of the Philadelphia Water Department's Saylor Grove Stormwater Wetland. (**a**) Image showing the wetland with a viewing platform and places for visitor's to site. (**b**) Signage at the viewing platform that educates visitors as to what they are observing, the ecological services this stormwater wetland provides, and the effects of stormwater and point and nonpoint source pollution. (**c**) Image showing the wetland looking back to the viewing platform

knowledge, skills, and approaches that minimize the negative impacts of too much information and; (3) people's desire to be needed, to make a difference, and to be treated respectfully.

The challenge of effective communication is twofold. First wetland specialists from both academia and industry need to listen to the public and understand what is important to them, but more importantly, work to include them as part of the decision-making process. The public understands recreational activities such as fishing, hunting, and tourism because it impacts their lives directly. Espousing the benefits of sediment/toxicant retention, nutrient removal/transformation, groundwater recharge/discharge, and floodflow alteration to the general public is nothing more than verbal ether. These services have no direct impact on their daily lives and they will not be interested. Such a conversation is more likely to generate questions about increased mosquito or tick populations; issues that impact them directly. This is not to say that sediment/toxicant retention or nutrient removal/transformation is not important. On the contrary, research on issues such as this is of the utmost importance, but wetland scientists need to significantly improve their ability to effectively communicate their findings to the general public. The importance of these results must be made relevant to the public in the context of positive or negative impacts to their daily lives. Public support is a powerful tool that can be used to enact rapid changes at societal and institutional levels. Minich (2011) provides a Landscape Architect's point of view for bridging the chasm between wetland scientists and the public. As she rightly points out, the success of a project generally rests on aesthetics and how the societal and ecological benefits are communicated to the public.

The second challenge that wetland scientists face is communicating the importance of all ecosystem services. The value of wetlands should be more strongly founded on an ecosystem approach, rather than the few ecosystem services that benefit the landowner or society. Landowners and society see tangible economic benefits from services such as recreation, fishing, and forestry. The value of any resource is based on how much the use of that resource improves the economic well-being of an individual or society (Pendelton 2009). However, services that are external to landowners such as sediment retention, storm surge protection, and toxicity reduction are not recognized as being important and are considered to have little or no value. While the public is generally well informed, subtle relationships between wetlands and storm surge (flooding of the coastal regions associated with hurricanes) or wetland loss and global warming are not necessarily intuitive. How many hurricanes like Katrina or Andrew will we need to endure before society understands that some of the damage incurred along our coastlines would have been prevented had the wetlands that had once occupied these regions been present? Scientists need to draw the line between the dots so that society understands these relationships clearly. The cleanup costs following such natural disasters are measureable and can be used to place an economic value on such services that have no apparent direct or immediate economic benefit to landowners.

The development of a wetland valuation system that is economically fair to landowners, while taking into account all ecosystem services will be difficult and certainly controversial. Wetland scientists together with economists and policy makers need to be united and aggressively promote the importance of the ecosystem services that are external to landowners and society and develop a wetland valuation system that includes all ecosystem services.

Towards an Adaptive Management Strategy

Although the multidisciplinary nature of the wetland sciences is recognized as being important, the reality is that most wetland sub-disciplines rarely communicate with one another. This is perhaps most apparent in wetland mitigation projects where wetlands are created for temporary and/or permanent impacts to wetlands within the project area. With few exceptions wetland mitigation is one of the final tasks in the construction sequence. As a result, mitigation wetlands are considered to be of minimal importance and inevitably poorly designed. As anticipated the financial resources dedicated for planning and implementation are limited and typically non-existent for post-construction maintenance activities. The result is that most of these wetlands end in failure.

The approach illustrated in Fig. 1.3 is perhaps best described as the "old way of doing things". Here the process is linear and prescribed. The wetlands are viewed as nothing more than fancy stormwater retention basins that are poorly designed and built with little consideration of the physical attributes of the site and input

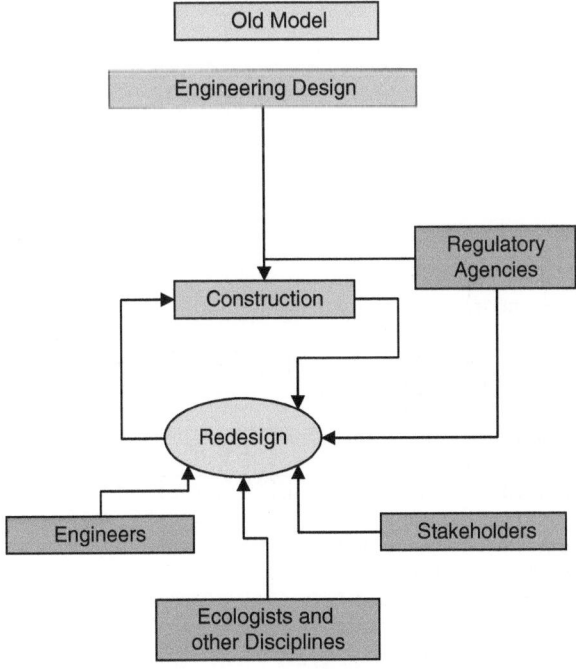

Fig. 1.3 Model showing the old approach to building wetlands. In most cases the wetland progressed from the design phase directly to construction. On occasion the regulatory agencies were involved, but their involvement was late in the process. As expected, many wetlands failed and it was at this stage that the stakeholders and other wetland scientists were brought in to correct the problems. The project was then re-designed to correct the flaws and moved back into the construction phase. Group discussion and consensus on sound approaches in the design are not a part of this model

from contributing disciplines and stakeholders. Throughout the process there is little opportunity for corrective actions. Inevitably these wetlands have limited functional capacity or fail completely. The problem is further exacerbated by the lack of consistent enforcement, which usually contributes to their abandonment. If sufficient funding remains or the activities are subject to enforcement, the wetland practitioners are then contracted to resolve the issues that could have been avoided had an adaptive management strategy been employed.

The adaptive management strategy presented in Fig. 1.4 is an example of a management strategy that has been developed for wetland construction projects and is intended here only to illustrate the concept. This approach brings together practitio-

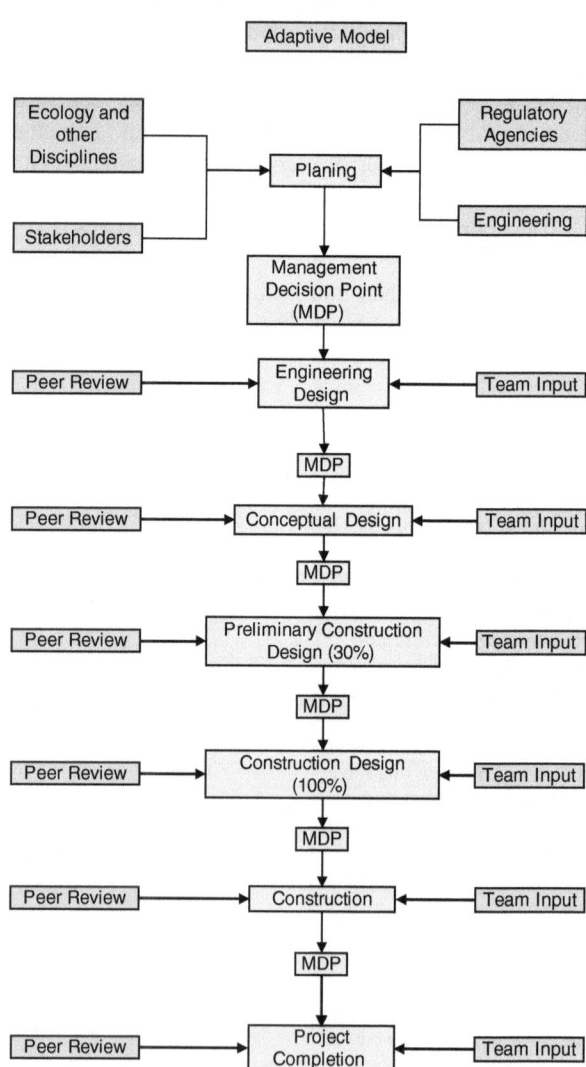

Fig. 1.4 The adaptive management strategy presented here provides continuous discussion and consensus between all team participants from the initial planning stage to completion. It allows for flaws in the design to be assessed and corrected before being implemented. This approach is becoming more accepted in the industry. As a result more created and restored wetlands are being deemed a success

ners and stakeholders from the wetland disciplines at the planning stage of the project. This approach provides the opportunity for all interested parties to gather the necessary baseline information on the site's physical attributes that will ultimately guide the planners towards an appropriate wetland design. More importantly, this approach provides the opportunity to identify and correct potential fatal flaws in the design and function of the wetland early in the planning process. As is the case in academia, this modified peer review process provides additional rigor to the decision making process.

The process is simple and illustrates a stepped approach from the initial design to the maintenance and monitoring phase. Such a strategy was employed during the design and construction of floating islands/wetlands on the Anacostia River in Washington, DC (see Hwang and LePage 2011). Each step requires peer review and group consensus before the process can move to the next level. This iterative approach provides the project team with the ability to identify potential problems early, discuss and develop solutions, and to make necessary adjustments to the project design continuously. The ability to react quickly is important, especially during the construction phase when the site conditions are continuously changing.

Conclusions

Wetlands are likely to become increasingly more prevalent in the urban landscape. As society's values come to embrace what wetlands can offer, sustainability will need to be redefined to include not only the typical ecological functions, but the social and economic. Functions external to landowners and society will inevitably become as important as those currently deemed to be important. Successful wetland projects both now and in the future will require an adaptive management strategy that integrates multiple disciplines in order to achieve the various goals and objectives expected by society. It is probably safe to say that as we move forward wetland managers will be tasked with maintaining and managing more wetlands with more input from stakeholders and with fewer resources.

References

Barbier EB, Acreman M, Knowler D (1997) Economic valuation of wetlands, a guide for policy makers and planners. Ramsar Convention Bureau, Gland, p 143

Boyd JW, Banzhaf S (2007) What are ecosystem services? The need for standardized environmental accounting units. Ecol Econ 63:616–626

Bradley GA (1995) Urban forest landscapes: integrating multidisciplinary perspective. University of Washington Press, Seattle, p 224

Brinson MM (1993) A hydrogeomorphic classification for wetlands. United States Army Corps of Engineers, Technical Report WRP-DE-4. Washington, DC, p 101. http://el.erdc.usace.army.mil/wetlands/wlpubs.html

Brinson MM (2011) Classification of wetlands. In: LePage BA (ed) Wetlands—integrating multidisciplinary concepts. Springer, Dordrecht

Cole CA, Kentula ME (2011) Monitoring and assessment—what to measure … and why. In: LePage BA (ed) Wetlands—integrating multidisciplinary concepts. Springer, Dordrecht

Connolly KD (2006) Wetlands law and policy: understanding Section 404. American Bar Association, Chicago, p 522

Council on Environmental Quality (2008) Conserving America's wetlands 2008: four years of partnering resulted in accomplishing the president's goal. The White House Council on Environmental Quality, Washington, DC, p 57

Dahl TE (2006) Status and trends of wetlands in the conterminous United States 1998 to 2004. United States Department of the Interior and United States Fish and Wildlife Service, Washington, DC, p 112

Dahl TE, Allord GJ (1996) History of wetlands in the conterminous United States. In: Fretwell JD, Williams JS, Redman PJ (eds) National water summary on wetland resources, United States Geological Survey, Water-Supply Paper 2425. United States Geological Survey, Washington, DC, pp 19–26. http://water.usgs.gov/nwsum/WSP2425/history.html

de Wet A, Williams CJ, Tomlinson J, Carlson Loy E (2011) Stream and sediment dynamics in response to Holocene landscape changes in Lancaster County, Pennsylvania. In: LePage BA (ed) Wetlands—integrating multidisciplinary concepts. Springer, Dordrecht

Dunnett N, Clayden A (2007) Rain gardens: managing water sustainably in the garden and designed landscape. Timber Press, Portland, p 188

Euliss NH, Smith LM, Wilcox DA, Browne BA (2008) Linking ecosystem processes with wetland management goals: charting a course for a sustainable future. Wetlands 28:553–562

Fisher B, Turner RK, Morling P (2009) Defining and classifying ecosystem services for decision making. Ecol Econ 68:643–653

France RL (2002) Wetland design: principles and practices for landscape architects and land use planners. W.W. Norton & Company, New York, p 160

Gardner RC, Davidson NC (2011) The Ramsar convention. In: LePage BA (ed) Wetlands—integrating multidisciplinary concepts. Springer, Dordrecht

Gwin SE, Kentula ME, Shaffer PW (1999) Evaluating the effects of wetland regulation through hydrogeomorphic classification and landscape profiles. Wetlands 19:477–489

Hwang L, LePage BA (2011) Floating islands: an alternative to urban wetlands. In: LePage BA (ed) Wetlands—integrating multidisciplinary concepts. Springer, Dordrecht

Kadlec RH, Wallace S (2008) Treatment wetlands, 2nd edn. CRC Press, Boca Raton, p 1048

Kaplan R (2011) Acknowledging and benefitting from multiple perspectives. In: LePage BA (ed) Wetlands—integrating multidisciplinary concepts. Springer, Dordrecht

Kaplan R, Kaplan S (2008) Bringing out the best in people: a psychological perspective. Conserv Biol 22:826–829

Keddy PA (2000) Wetland ecology: Principles and conservation. Cambridge University Press, New York, NY, p 614

Kiesecker JM, Copeland H, Pocewicz A, McKenney B (2010) Development by design: blending landscape-level planning with the mitigation hierarchy. Front Ecol Environ 5:261–266

LePage BA (2007a) Reconstructing wetland vegetation using pollen and the plant fossil record for wetland creation and enhancement. Society of Wetland Scientists Annual International Meeting, Sacramento, 10–15 June 2007

LePage BA (2007b) Use of the pollen and the plant fossils for wetland creation, restoration and enhancement. National watershed-wide strategies to maximize wetland ecological and social services. Association of State Wetland Managers, Inc. 28–29 Aug 2007

Magee TK, Kentula ME (2005) Response of wetland plant species to hydrologic conditions. Wetlands Ecol Manage 13:163–181

Magee TK, Ernst TL, Kentula ME, Dwire KA (1999) Floristic comparison of freshwater wetlands in an urbanizing environment. Wetlands 19:517–534

McInnis RJ (2011) Managing wetlands for multifunctional benefits. In: LePage BA (ed) Wetlands—integrating multidisciplinary concepts. Springer, Dordrecht

Middleton B (2011) Multidisciplinary approaches to climate change questions. In: LePage BA (ed) Wetlands—integrating multidisciplinary concepts. Springer, Dordrecht

Millennium Ecosystem Assessment (2005) Ecosystems and human well-being: wetlands and water synthesis: a report of the Millennium Ecosystem Assessment. World Resources Institute, Washington, DC, p 68

Minich N (2011) The role of landscape architects and wetlands. In: LePage BA (ed) Wetlands—integrating multidisciplinary concepts. Springer, Dordrecht

Mitsch WJ, Gosselink JG (1986) Wetlands. Van Nostrand Reinhold, New York, p 539

Murray B, Jenkins A, Kramer R, Faulkner SP (2009) Valuing ecosystem services from wetlands in the Mississippi alluvial valley. Ecosystem Services Series NI R 09-02. Nicholas Institute for Environmental Policy Solutions, Duke University, Durham, North Carolina, p 43

Nahlik AM, Kentula ME, Ringold P, Landers D, Weber M (2010) Potential indicators of ecosystem services in wetlands. 2010 Society of Wetland Scientists Annual Meeting, Salt Lake City, 27 June–2 July 2010

Nowak DJ, Twardus D, Scott CT (2000) Proposal for urban forest health monitoring in the US. Society of American Foresters National Conference Proceedings, Washington, DC, pp 178–183

Nyman JA (2011) Ecological functions of wetlands. In: LePage BA (ed) Wetlands—integrating multidisciplinary concepts. Springer, Dordrecht

Pendelton LH (2009) The economic and market value of coasts and estuaries: what's at stake? Restore America's Estuaries, Arlington, p 175

Ramsar (2010) http://www.ramsar.org. Accessed 4 June 2010

Robertson M, Hough P (2011) Wetlands regulation: the case of mitigation under section 404 of the clean water act. In: LePage BA (ed) Wetlands—integrating multidisciplinary concepts. Springer, Dordrecht

United States Environmental Protection Agency (2009) Valuing the protection of ecological systems and services: a report of the EPA science advisory board, EPA-SAB-09-012. United States Environmental Protection Agency, Washington, DC, p 121

United States Environmental Protection Agency (2010) http://www.epa.gov/ecology/publications.htm

Vymazal J, Kröpfelová L (2008) Wastewater treatment in constructed wetlands with horizontal sub-surface flow I. Springer, New York, p 566

Williams CJ (2011) A paleoecological perspective on wetland restoration. In: LePage BA (ed) Wetlands—integrating multidisciplinary concepts. Springer, Dordrecht

World Bank (2007) Global economic prospects 2007: managing the next wave of globalization. The International Bank for Reconstruction and Development, Washington, DC, p 180

Zeff ML (2011) The necessity for multi-disciplinary approaches to wetland design and adaptive management: the case of wetland channels. In: LePage BA (ed) Wetlands—integrating multidisciplinary concepts. Springer, Dordrecht

Chapter 2
The Necessity for Multidisciplinary Approaches to Wetland Design and Adaptive Management: The Case of Wetland Channels

Marjorie L. Zeff

Abstract The need for a multidisciplinary approach to the design and adaptive management of constructed wetlands is illustrated by case examples of channel form and function in a variety of wetland types. Channels in wetland systems are typically viewed simply as conduits of water inflow and outflow. However, there are dynamic interrelationships amongst vegetation, hydrology/hydraulics, and substrate in wetland channel systems that demand a more holistic approach to wetland management that considers the disciplines of biology, engineering, and sedimentary geology. Recognition of the inter-dependence of the biologic, hydrologic, and geomorphic components of channelized flow in wetland systems is critical to the successful design of self-sustaining constructed wetlands.

For many wetlands, channelized flow is the predominant source of hydrology necessary to sustain the system. In other wetlands, channelized flow may be equally as important as groundwater and overland sheet flow as a water source. And, as wetlands can be defined by hydrology, vegetation, and soils, so are the form and function of wetland channels defined by the interrelationships of hydrology, vegetation, and soils.

In recent years, the practice of stream restoration has identified the need to integrate the sciences of engineering, biology, and geology in the application of fluvial geomorphology. This has been motivated by the less than stellar success rate for stream restoration projects. Too often the multidisciplinary nature of the natural system is lost by the dominance of one of these technical fields in the design of the restoration site. This can also be said of the design of channels in the practice of wetland construction for the purposes of restoration, enhancement, and creation.

Wetland channels serve as excellent examples of how important it is to engage multiple disciplines in the study of wetlands and the application of wetland science. On the surface, it might appear that channelized flows in wetlands are purely

M. L. Zeff (✉)
URS Corporation, 335 Commerce Drive, Fort Washington, PA 19034, USA
e-mail: marjorie_zeff@urscorp.com

B. A. LePage (ed.), *Wetlands,*
DOI 10.1007/978-94-007-0551-7_2, © Springer Science+Business Media B.V. 2011

a matter for hydraulic engineers charged with designing a conduit to convey water. However, this could not be further from the truth. Channels in a wetland constrain the development of its biological character. Understanding how the bio-, hydro-, and geomorphic components of a channelized system in a wetland are inter-dependent and unique is essential if we are to design self-sustaining constructed wetland systems. When present, channels are lynch-pins in the holistic integrity of a wetland system.

The focus of this chapter will be to take a look at wetland channel morphology in a variety of wetland types to illustrate the inter-related contributions of geology, engineering, and biology to the understanding of wetlands. It is not intended to be a comprehensive synopsis of wetland channel knowledge, nor a literature review of wetland channel studies. The references cited in this chapter, in combination, will accomplish that. Its objective is to introduce the reader to the wide range of multidisciplinary linkages in wetland channel morphologies and functions so as to emphasize the need to incorporate a multidisciplinary approach to the design and adaptive management of wetland creation, enhancement, and restoration sites.

Many terms and classification systems are used to describe and define wetlands—swamp, marsh, bog, fen, peatland, mire, moor, muskeg, bottomland, wet prairie, reedswamp, wet meadow, slough, pothole, playa (Mitsch and Gosselink 2000). And, many terms have been used to refer to linear features of water conveyance in wetlands—streams, creeks, channels, and sloughs. The original terms used in the studies cited are maintained here. No attempt has been made to re-assign wetland types and terms to a single classification system.

Channels and Feedback Mechanisms

Wetland landforms display hydrology, vegetation, and soil characteristics distinguishable from non-wetland systems. Aside from land preservation or the regulation of activities that could directly or indirectly affect wetlands, wetland protection in the United States today is largely achieved by replacing wetland functions lost or diminished by some action. Channels in wetland functional analyses are typically viewed as conduits of water inflow and outflow. Similarly, when designing or monitoring channels in constructed wetlands, the focus is typically to size the conveyance of computed quantities of water into and out of the system. Channels however, play a role beyond serving as conduits delivering water and carrying it away.

When it comes to channels, there is a temptation to view alluvial rivers and their floodplains as templates for wetland channels and the marshes or swamps they pass through. However, as stated by Jurmu (2002, p. 832): "If wetlands are unique biologically, have distinct hydrology, and function unlike other environments, it follows that factors affecting streams (their function and characteristics) might also be distinct and create different morphological features." To this can be added that there is considerable variety in wetland types, hence, one can expect the variety in form and function of wetland channels to be similarly vast.

There is also a tendency to view wetland systems as static, and not evolutionary. Channels maintain the existing integrity of the system while simultaneously advancing changes which will lead either to continued sustainability or not. The feedback influences of biology, geology, and hydrology on one another create a changing system with time. When establishing protocols of adaptive management for constructed wetlands it is very important to understand how channel processes influence the temporal development of the wetland as a whole.

How do wetland properties influence channel form and how does channel form affect overall wetland hydrology? How does this vary among different wetland types? The scientific literature reveals a dynamic interrelated feedback amongst vegetation (biology), hydrology/hydraulics (engineering), and substrate (sedimentary geology). The fact of this inter-dependence demands a holistic approach to wetland management in general, and wetland construction in particular (Fig. 2.1). There is a need to integrate a developing school of thought within the disciplines of geomorphology and ecology that recognizes the imperative to integrate the physical and biological.

The recognition of the importance of the reciprocal interactions and adjustments of biotic (organisms and communities) and abiotic (form and process) components of our planet has been fundamental to the development of modern science in the nineteenth century (Corenblit et al. 2008). Landscapes develop as by-products of the feedback mechanisms between organisms and their habitats which are simultaneously dependent and controlling factors.

Geomorphic-biological feedback, the interactions between the geomorphic and ecological components of landscape, has often been viewed as independent pro-

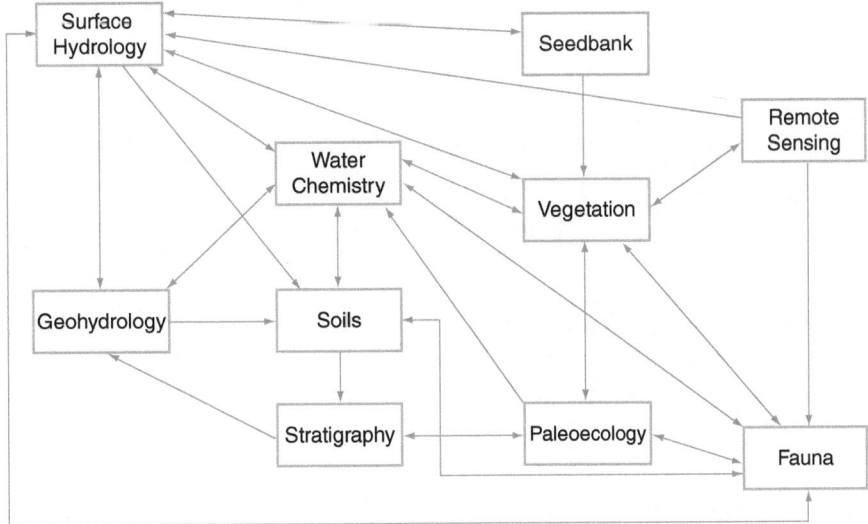

Fig. 2.1 Arrow directions indicate how different wetland disciplines can inform others. (After Wilcox 1987)

cesses operating in one direction (Stallins 2006). That is, the scientific literature provides many instances where, in one direction, geomorphic process and landscape are shown to influence biota or, in the other direction, the biota is shown to affect geomorphic process and landscape. This unidirectional precept, however, is being replaced with more complex, non-linear developmental and evolutionary theories wherein form and process evolves in accordance with biologic evolution, in the long term, and with ecological succession in the shorter. There is a long- and short-term developmental linkage and a cumulative feedback, a sort of bio-geomorphologic inheritance or memory.

Multiple Influences on Channel Morphology—Case Examples

When referring to channel morphology we must consider plan form and cross-section shape (Figs. 2.2 and 2.3). Fauna and flora are documented to have an influence on channel form and process, and vice versa. Case examples are provided below to demonstrate the wide range of channel bio-geomorphic and hydro-geomorphic feedback mechanisms to be found in many wetland types. It is not intended to be

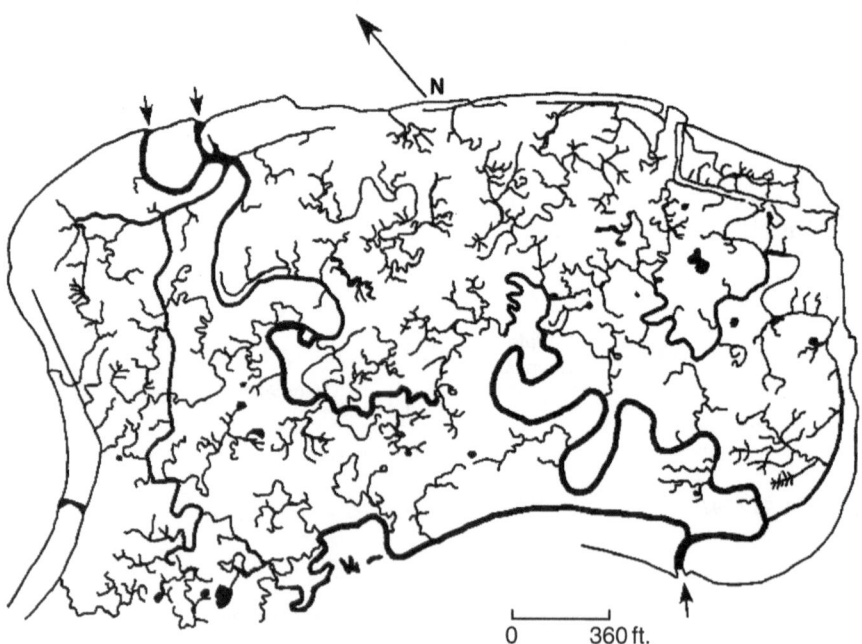

Fig. 2.2 Channel distribution in natural tidal marshes illustrating the complex plan form morphometry of a channel system in a tidal marsh of San Francisco Bay. Varying morphometric parameters include order, sinuosity, drainage density, and junction angle. Arrows indicate major channel inlets/outlets. (Redrawn and modified from Pestrong [1965])

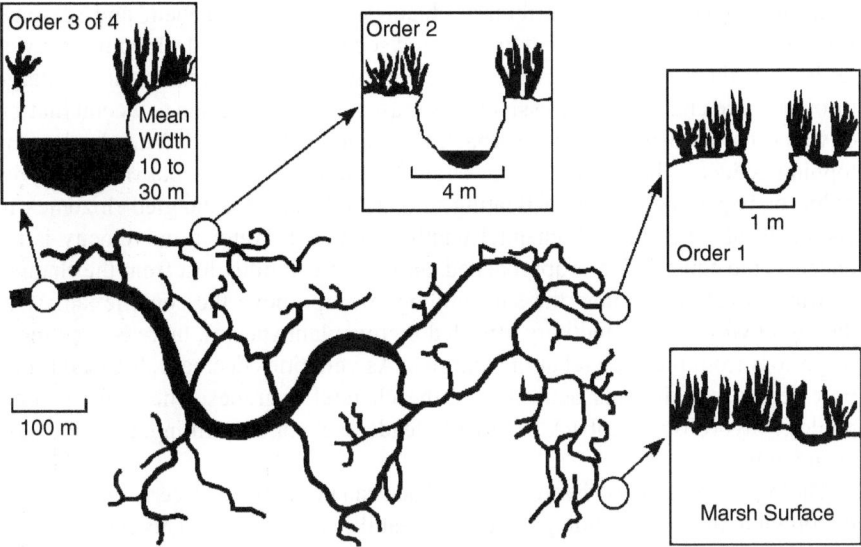

Fig. 2.3 Creek network illustrating how channel cross-sectional morphometry (e.g., width:depth ratio) and hydrology vary with position in the system. The insets illustrate the typical profiles and widths of the creeks and water levels at low tide. The third and fourth order creeks retain most of the water volume at low tide, while the second order creeks retain shallow pools at low tide. The first order creeks and marshes drain completely at low tide. (Re-drawn and modified from Williams and Desmond [2001])

exhaustive, but rather to provide support for the approach to wetland construction design and adaptive management to be discussed subsequently.

Vegetation has been the focus of most bio-geomorphic feedback investigations with respect to wetland channels, particularly the effect on flow of increased bank strength and hydraulic roughness from vegetation along channel margins. Jurmu (2002) suggests this might lead to the low width:depth ratios (relatively narrow and deep) of stream reaches passing through palustrine wetlands (emergent, scrub–shrub, forested) in Connecticut, Indiana, and Wisconsin. In these cases, channel shape adjusts to flow variability by bed erosion and deepening rather than lateral channel migration that is prevalent in alluvial channels.

In tidal San Francisco Bay, channel sinuosity is higher in salt marshes, where the vegetation is a dominant feature, than on adjacent unvegetated tidal mudflats (Pestrong 1972). In New Jersey, USA higher channel sinuosity has been found in tidal salt marshes with dense and extensive root systems in peaty substrates than in tidal freshwater marshes with sparse root systems in muddy substrates (Garofalo 1980). Channel morphology is influenced more by hydrodynamic factors in the freshwater tidal marshes while vegetation is more important in the salt marshes.

In southern Africa's largest wetland, root density and root attachment to a peaty substrate influences channel form and function in a very different way (McCarthy and Ellery 1997; Ellery et al. 2003; Tooth and McCarthy 2004). In Botswana, the

Okavango River and its distributary channels support permanent and seasonal floodplain swamps of emergent grasses and sedges by overbank flooding and water leakage through channel margins. In contrast to alluvial streams, the discharge through these channels progressively decreases downstream due to a combination of water loss to distributaries, overbank flow, and bank leakage. When bank overtopping reduces channel discharge enough to allow *Cyperus papyrus* (papyrus sedge or paper reed), a semi-floating unattached mat of entangled rhizomes to encroach into the channel, channel width is reduced. Water flow velocity is reduced as the constriction is approached, and water is rapidly lost from the channel as water levels rise above the surrounding swamp above the constriction. With the high hydraulic conductivity afforded by this plant species, there is even more water loss from the channel at the margins. As velocities decrease, bedload transport declines, sediments deposit, and the channel aggrades. The biology–morphology–hydrology feedback eventually leads to channel in-filling, avulsion, and abandonment.

Biology–morphology–hydrology feedback can be subtle to discern, but is no less important to the understanding of the sustainability of a wetland. This can be seen in the case of channels through peatland fens in Wisconsin and Canada.

Watters and Stanley (2007) investigated the cross-section and plan form morphology of a stream flowing through an extensive fen in Wisconsin, USA. Over 90% of total stream flow is groundwater base flow, thus, channel discharge variability is low and overbank flooding is rare. Vegetation near the stream is mostly hummock-forming sedges. An organic channel bed substrate prevails in the fen; an inorganic, mineral substrate outside the fen. An interesting characteristic of the fen channel cross-section is a shallow side with loose organic sediments (highly decomposed with 25–50% organic content) and a deeper side with firm peat (90% organic content with limited decomposition). The shallow sides are zones of groundwater discharge. Thus, peat dynamics, dominated by groundwater hydrology, dictate plant decomposition and peat quantity and quality (fiber content, susceptibility to decomposition, and bulk density). All this is linked back to channel morphology.

One can extend the biofeedback concept through time whereby the biologic mechanisms affecting channel form are past processes. This is exemplified by the case of distributary channels in the Cumberland Marshes in Canada (Smith and Perez-Arlucca 2004). Here, channels are incising through peats produced by old fens. Fen peatlands originally occupying alluvial floodplains were converted to shallow basins after being flooded by the avulsion of a main channel. Distributary channel networks developed over a wedge of avulsion sediment that covered the peats. Downcutting channels may eventually encounter the pre-avulsive peat. Channels with peat bottoms tend to have rectangular cross-sections, higher width:depth ratios, and higher average:maximum depth ratios than channels that have not yet reached the peat, or have completely eroded through it. This suggests that when encountered, the peat promotes the accommodation of increasing discharges by enlarging through channel widening rather than deepening. This is a biologic influence on channel shape that is temporally disjointed from current processes.

An informative way to add flow to the equation to see how channel shape adjusts to changing flow regimes is to assess hydraulic geometry, relating changes in discharge to rates of change in channel width, depth, and velocity. The hydraulic geometry relationships for a wide range of tidal and non-tidal wetland types demonstrate the variety of channel responses to hydrologic forces (Myrick and Leopold 1963; Zeff 1988; Leopold et al. 1993; Tooth and McCarthy 2004; Watters and Stanley 2007; Diefenderfer et al. 2008; Nanson et al. 2010).

Nanson et al. (2010) identify an interesting hydraulic geometry relationship in peatland swamps of Australia. Channel banks are nearly vertical with high bank strength provided by grass- and tussock-rooted peat. Channels remain relatively narrow and deep in these swamps, reflected in low width:depth values. However, the hydraulic geometry relations indicate that these channels accommodate increases in discharge by increasing flow velocity rather than adjusting channel dimensions. Bankfull flows are frequent enough to maintain high enough water tables to support the wetland vegetation, however, bankfull flows are quickly moved through the channels and overbank flooding is rare. Vertical growth of the wetland is limited and linked to channel depth.

Channels in Wetland Design and Adaptive Management

It should be evident from the few cases cited above that, when designing channels for constructed wetlands, one needs to consider more than the hydrology/hydraulics concern for sizing to convey predicted design discharges. As illustrated in these studies, channels will accommodate changes in flow regime by adjusting shape, or velocity. And, the channel response to process alterations will be unique to the biogeomorphology and hydrogeomorphology of the particular wetland.

Predicting how these adjustments will impact the short-term success and long-term sustainability of a constructed wetland requires a robust multidisciplinary understanding and application of site-specific feedback mechanisms of biology, hydrology/hydraulics, and sedimentary geology. And, to establish when adaptive management is appropriate, the short-term and long-term integrity of the entire wetland system needs to be evaluated with respect to intrinsic evolution and extrinsic forces. For example, do we need to interfere with sedimentation in a channel if infilling and avulsion is a process-response necessary for continued existence of the wetland system? What can be done to protect a restored salt marsh from drowning if vertical accretion is not keeping pace with sea level rise?

The multidisciplinary approach needed is really an interdisciplinary approach. Wilcox (1987) noted that the scientific studies of various disciplines regularly involved in wetland research are often narrow in focus and there is a danger in extrapolating these limited scopes to broader wetland issues. He encouraged the collective interpretation of data from multiple disciplines. This holds true for constructed wetland and post-construction management as illustrated by the case of the wetland channels.

References

Corenblit D, Gurnell AM, Steiger J, Tabacchi E (2008) Reciprocal adjustments between landforms and living organisms: extended geomorphic evolutionary insights. Cateña 73:261–273

Diefenderfer HL, Coleman AM, Borde AB, Sinks IA (2008) Hydraulic geometry and microtopography of tidal freshwater forested wetlands and implications for restoration, Columbia River, USA. Ecohydrol Hydrobiol 8:339–361

Ellery WN, McCarthy TS, Smith ND (2003) Vegetation, hydrology, and sedimentation patterns on the major distributary system of the Okavango Fan, Botswana. Wetlands 23:257–375

Garofalo D (1980) The influence of wetland vegetation on tidal stream channel migration and morphology. Estuaries 3:258–270

Jurmu MC (2002) A morphological comparison of narrow, low-gradient streams traversing wetland environments to alluvial streams. Environ Manage 30:831–856

Leopold LB, Collins JN, Collins LM (1993) Hydrology of some tidal channels in estuarine marshland near San Francisco. Cateña 20:469–493

McCarthy TS, Ellery WN (1997) The fluvial dynamics of the Maunachira Channel system, northeastern Okavango Swamps, Botswana. Water SA 23:115–125

Mitsch WJ, Gosselink JC (2000) Wetlands, 3rd edn. Wiley, New York, p 920

Myrick RM, Leopold LB (1963) Hydraulic geometry of a small tidal estuary. United States Geological Survey, Professional Paper 422-B:1–18

Nanson RA, Nanson GC, Huang HQ (2010) The hydraulic geometry of narrow and deep channels; evidence for flow optimisation and controlled peatland growth. Geomorphology 117:143–154

Pestrong R (1965) The development of drainage patterns in tidal marshes. Stanford University Publications in Earth Science 10:1–878

Pestrong R (1972) Tidal-flat sedimentation at Cooley Landing, southeast San Francisco Bay. Sed Geol 8:251–288

Smith ND, Perez-Arlucca M (2004) Effects of peat on the shapes of alluvial channels: examples from the Cumberland Marshes, Saskatchewan, Canada. Geomorphology 61:323–335

Stallins JA (2006) Geomorphology and ecology: unifying themes for complex systems in biogeomorphology. Geomorphology 77:207–216

Tooth S, McCarthy TS (2004) Controls on the transition from meandering to straight channels in the wetlands of the Okavango Delta, Botswana. Earth Surf Process Landforms 29:1627–1649

Watters JR, Stanley EH (2007) Stream channels in peatlands: the role of biological processes in controlling channel form. Geomorphology 89:97–110

Wilcox DA (1987) A model for assessing interdisciplinary approaches to wetland research. Wetlands 7:39–49

Williams GD, Desmond JS (2001) Restoring assemblages of invertebrates and fishes. In: Zedler JB (ed) Handbook for restoring tidal wetlands. CRC Press, Boca Raton, pp 235–269

Zeff ML (1988) Sedimentation in a salt marsh-tidal channel system, southern New Jersey. Mar Geol 82:33–48

Chapter 3
Stream and Sediment Dynamics in Response to Holocene Landscape Changes in Lancaster County, Pennsylvania

Andrew deWet, Christopher J. Williams, Jaime Tomlinson
and Erin Carlson Loy

Abstract Sediment pollution is one of the most important contributors to the degradation of the Chesapeake Bay. The Susquehanna River is one of the main sources of sediment to the Bay and this source could dramatically increase in the future. The reservoirs on the Susquehanna River are almost at sediment-storage capacity. The lowest reservoir (Conowingo) could be at capacity within a few decades (Langland and Hainly 1997). Once this occurs little sediment will be trapped and prevented from entering the Bay. Sediment continues to be supplied to the river system from across the watershed. Understanding the sources of this sediment, as well as the processes contributing to the continued flow of sediment, are crucial if efforts to restore the Bay's ecosystem are to be successful.

Our study has focused on one of the sub-basins of the Susquehanna River located in Lancaster County, Pennsylvania. We have examined the spatial and temporal distribution of sediment in the stream channel and floodplain and linked this information to changes in the landscape over the last several thousand years. Typically we see coarse-grained channel lag deposits formed by lateral migration of streams which are overlain by predominately fine-grained organic-rich (including large logs) sediment that accumulated slowly over several thousand years (based on radiocarbon ages). Deforestation resulting from European occupation beginning in the early 1700's caused a significant increase in the rate of sediment deposition in the floodplain and a decrease in the organic content of the sediment. Debris flows from the surrounding hillslopes resulted in matrix-supported conglomerates within the floodplain sediments. Modifications to the stream channels to facilitate agriculture were also common. More recent landscape changes resulting from suburbanization have resulted in complex spatial and temporal responses in the stream system. Initially, sediment deposition in the channels increased, however changes in land development practices have resulted in increased peak stream flows and the remobilization of the sediment from the stream channels, streambanks, and

A. deWet (✉)
Department of Earth and Environment, Franklin and Marshall College,
Lancaster, PA 17603, USA
e-mail: andy.dewet@fandm.edu

B. A. LePage (ed.), *Wetlands,*
DOI 10.1007/978-94-007-0551-7_3, © Springer Science+Business Media B.V. 2011

floodplains. Most streams are incising and/or widening their channels causing high sediment yields. The resulting sediment pollutes the local streams and contributes to ongoing degradation of the Chesapeake Bay.

Stream channels and adjacent riparian areas are dynamic environments that continually adjust to the natural and anthropogenic changes within a watershed. Specific channel and floodplain morphologies result from the flow of water and the movement of sediment and other transported materials within the channel and between the channel and floodplain. As noted by Naiman et al. (2002), this relationship is fundamental to the creation, maintenance, and ecological integrity of fluvial ecosystems. Longer-term (greater than 1,000 years) climate change that results in different temperature and precipitation regimes alter the supply of sediment and water to the stream channels, which in turn drives changes in channel and floodplain morphology (Johnstone et al. 2006; Macklin et al. 2006). On the other hand, human activity has directly and indirectly modified the natural water flow regimes and altered the dynamic equilibrium between the flow of water and the movement of sediment (Poff et al. 1997; Vörösmarty and Sahagian 2000; Walter and Merritts 2008). However, these two forcing agents need not operate in isolation and in fact, human induced landuse change can act as either a positive or negative feedback mechanism on climate change related effects on stream sediment dynamics (Macklin and Lewin 2003). It is beneficial to understand the role that these two agents of change play in shaping the sediment dynamics and geomorphology of present day stream channels. From a purely practical standpoint, understanding the response rate and behavior of fluvial ecosystems to landuse change would enhance our ability to model and manage (including restoration, enhancement and creation efforts) stream, riparian, and wetland ecosystems in the future.

Our study has focused on one of the sub-basins of the Susquehanna River located in Lancaster County, Pennsylvania, USA. We have examined the spatial and temporal distribution of sediment in the stream channel and floodplain and linked this information to changes in the landscape over the last several thousand years. Typically we see coarse-grained channel lag deposits formed by the lateral migration of streams that are overlain by predominately fine-grained organic-rich (including large logs) sediment that accumulated slowly over several thousand years (based on radiocarbon ages). Deforestation resulting from European occupation beginning in the early 1700s caused a significant increase in the rate of sediment deposition in the floodplain and a decrease in the organic content of the sediment. Debris flows from the surrounding hillslopes resulted in matrix-supported conglomerate deposits within the floodplain. Structural modifications to the stream channels to facilitate agricultural practices were also common. More recent landscape changes resulting from suburbanization have resulted in complex spatial and temporal responses in the watershed's stream system. Initially, sediment deposition in the channels increased, however changes in land development practices resulted in increased peak stream flows and the remobilization of the sediment from the stream channels, streambanks, and floodplains. Most streams are incising and/or widening their

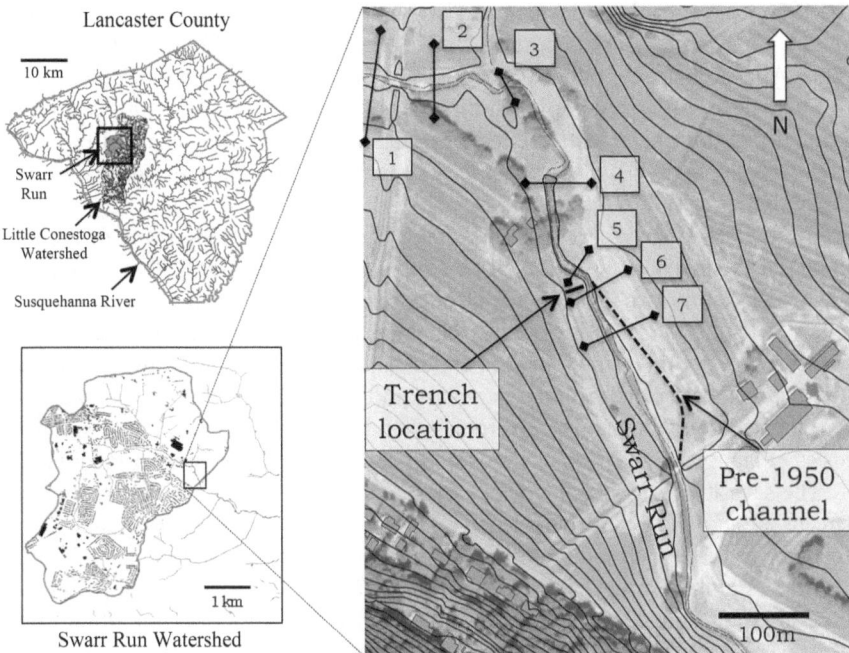

Fig. 3.1 Location map of the study reach of Swarr Run, Lancaster County, Pennsylvania. Topographic cross-sections are numbered *1–7*. The trench is located on the south-west side of the channel and the position of the original pre-1950 channel along the study reach is located to the east of the existing channel

channels causing high sediment yields. The resulting sediment pollutes the local streams and contributes to ongoing degradation of the Chesapeake Bay (Langland and Hainly 1997).

In this chapter we examine how changes in land cover and landuse influenced the erosion and deposition of sediment in a small second order (fed by a few small first order tributaries) stream called Swarr Run (Fig. 3.1). The Swarr Run Watershed is 18.6 km^2 (7.2 square miles) and is part of the larger Conestoga Watershed, which is approximately 1,218 km^2 in size (470 square miles) and located in Lancaster County, Pennsylvania (Loper and Davis 1998; Langland et al. 2001). The Conestoga Watershed occupies only 0.76% of the Susquehanna drainage basin, which is approximately 168,349 km^2 in size (65,000 square miles), yet it contributes disproportionally high, discharge-weighted sediment concentrations to the Chesapeake Bay Watershed (Gellis et al. 2004). Given the relatively recent shift towards state-by-state tributary scale restoration efforts of degraded streams in the Chesapeake Bay Watershed (Hassett et al. 2005) Swarr Run is a good representation of the complex challenges associated with reducing the sediment and nutrient loads to the Chesapeake Bay. Climate changes since the last ice age, post-European settlement deforestation and agriculture, followed by rapid suburbanization since the 1970s, have resulted in widespread changes to the stream processes and downstream environments in the eastern United States (Trimble 1997; Paul and Meyer 2001; Allan 2004; Leopold et al. 2005). Thus, the present day geomorphic features of Swarr Run reflect a complex and perhaps unique history superimposed on the general post-European settlement deforestation—agriculture—channel modification—development landuse-trajectory of many locations in the North American Piedmont.

By the late 1990s, several reaches of Swarr Run were likely some of the most impaired stream channels in Lancaster County (Tomlinson 2003). The stream illustrated many of the problems that confront stream restoration activities in developed and developing landscapes. Locally, channelization and unrestricted farm animal access to the stream resulted in accelerated bank erosion and instability, a wide and sediment choked channel, and a degraded pool-riffle structure (Bayrd et al. 2002). An elevated floodplain and steep eroding banks reflected decades and possibly centuries of landscape changes, channel modifications, and poor farming practices.

The severity of the degraded state of the stream prompted efforts to restore the stream through the Pennsylvania Growing Greener Program; a state grant program focused on watershed restoration and protection. During the summer of 2003 a stream restoration project targeted 1.1 km (0.7 miles) of the most degraded section of Swarr Run. The excavation activity associated with the restoration efforts exposed a deep sequence of sediments providing an opportunity to examine the stratigraphic record of floodplain sediment deposition and erosion surfaces at this site dating back to the early Holocene (*ca.* 10,000 Before Present [BP]). We used these data to better understand the response of the stream system to both natural and anthropogenic changes that have occurred in this part of the watershed. Moreover, by understanding how such ecosystems responded to past changes, it is possible to adopt and implement restoration practices that will not only restore the functioning of such ecosystems but implement changes that are consistent with how the stream functioned in the past.

Fieldwork/Observations

Several topographic cross-sections were measured across the floodplain (Figs. 3.1 and 3.3). Sediments in the Swarr Run floodplain were examined along the numerous exposed banks along the stream and in a trench excavated across the floodplain (Figs. 3.1 and 3.2). During the restoration work, a 23 m long trench was excavated perpendicular to the existing stream channel and extended across the floodplain from the upland slope to within a few meters of the channel. The trench varied from 1 to 3 m deep. The stratigraphy of the sediments was determined using physical characteristics of the sediment including the Munsell soil color, grain size, and texture.

Sediment samples were collected for grain-size, geochemical, loss on ignition (LOI), and micro- and macrofossil analyses. Major stratigraphic units at five loca-

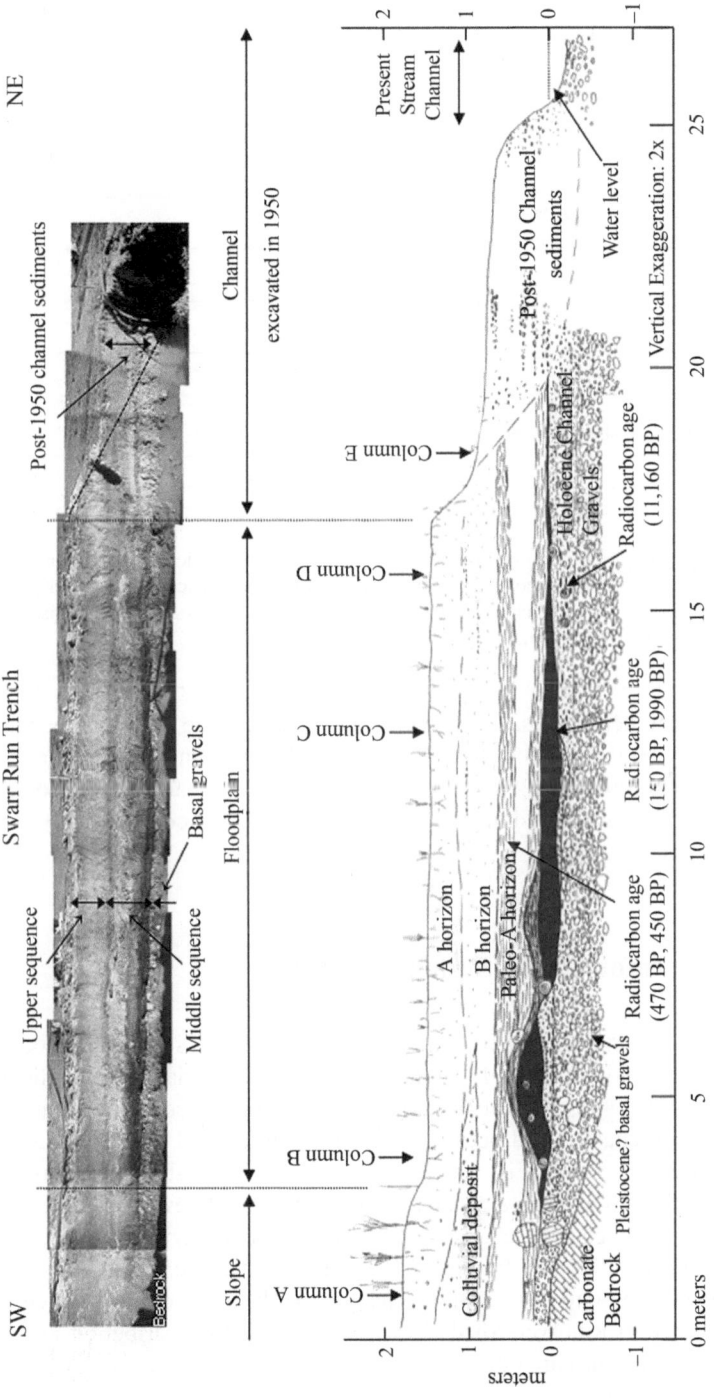

Fig. 3.2 Details of the floodplain sediments exposed in the trench excavated across the floodplain to the south-west of the study reach (see Fig. 3.1). The present stream channel is located to the right (north-east) and the slope to the left (south-west). The bedrock, basal gravels, middle sequence, upper sequence, and the post-1950 channel sediments are identified

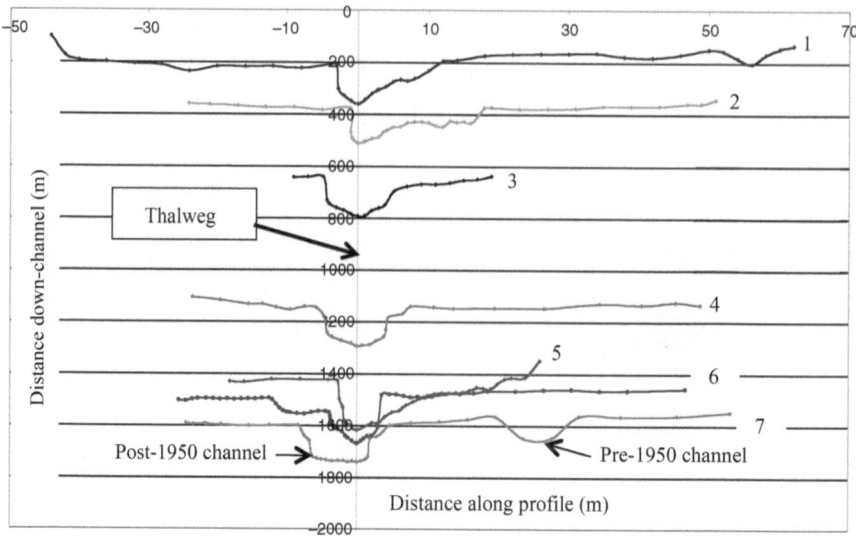

Fig. 3.3 Topographic cross-sections measured across the floodplain along the study reach (see Fig. 3.1). The cross-sections are positioned relative to the thalweg and according to their location downstream. Section *6* is located closest to the trench and section *7* shows the present channel and the location of the pre-1950 channel

tions along the trench were sampled with a hand trowel. The sample sizes ranged from over 2 kg for the coarsest units to 200 g for the very fine units. The color of the sediment was measured *in situ* using a Munsell Color Chart. Buried wood and charcoal identified in the sediments were collected for radiocarbon analysis and taxonomic identification.

Geographic Information System Analysis

Field observations indicated that Swarr Run had undergone significant modifications since early colonial times. Examination of historical aerial photographs from 1936, 1947, 1951, 1958, 1964, 1974, 1983, 1992, and 2001 confirmed that in the 1950s the stream channel had been extensively modified and completely relocated along at least one reach (Fig. 3.1). Additionally, the watershed upstream of the field site experienced extensive residential and commercial development since the 1930s. These watershed-scale changes influenced the stream dynamics and the depositional/erosive condition of the stream. We constructed a Geographic Information System (GIS) database to document the changes that have occurred in the channel and across the Swarr Run Watershed. We used changes in the distribution, area, and number of buildings through time (1936, 1947, 1964, 1974, 1983, 1992, and 2001) as a measure of the residential/commercial development that has occurred in the watershed. This information was then correlated with the field and GIS observations of the changes in the channel and the depositional/erosional record that are preserved in these valley sediments.

Laboratory Analyses

Grain Size Analysis

Field samples that weighed between 0.2 and 2 kg were collected. All of the samples were initially dried and disaggregated. For the coarse (gravel) samples, the whole sample was dry sieved. The less than 2 mm (−1 phi) fraction was then split and dry sieved down to 0.125 mm (3.0 phi). The finer grained samples were split and between 30 and 180 g of sample were dry sieved down to 0.125 mm (3.0 phi). The less than 0.125 mm fraction was split and 3 g of sample was analyzed using a Sedigraph at the Graduate College of Marine Studies, University of Delaware.

Loss on Ignition

LOI was measured to estimate the organic and inorganic carbon content of the less than 2 mm size fraction of selected samples from the trench. The samples were dried at 100°C for 2 h and weighed. The samples were then heated to 450°C for 2 h and re-weighed to determine the mass loss of the organic carbon fraction. Each sample was then heated for an additional 2 h at 850°C and then re-weighed to determine the mass loss due to the inorganic carbon fraction.

Radiocarbon Dating

Four samples of buried wood and one sample of charcoal were collected from Swarr Run and carbon dated. Three samples were collected from the exposed streambank 10 m upstream of the trench site (SW1, SW2, and SW3) and two samples were collected from the trench site (SRTC1 and SRTS17) (Fig. 3.1 and Table 3.1). All radiocarbon analyses were carried out by Beta Analytic Radiocarbon Dating Laboratory. Samples were analyzed by synthesizing sample carbon to benzene (92% carbon), measuring for ^{14}C content in a scintillation spectrometer, and then calculating for radiocarbon age. All ages are expressed as calibrated ages based on the INTCAL 98 database (Stuiver et al. 1998).

Table 3.1 Location and summary of the ^{14}C ages from wood and charcoal sampled from the flood-plain sediments, Swarr Run, Lancaster County, Pennsylvania

Sample No.	Latitude (°N)	Longitude (°W)	Material dated	^{14}C age (conventional)	1σ (years BP)	Calibrated age (years BP)	+/−2σ (years BP)
SRTC1	40°04′52.853″	076°22′45.609″	Char.	360	40	450	510–310
SW2	40°04′53.352″	076°22′45.532″	Wood	380	60	470	530–300
SW1	40°04′53.352″	076°22′45.532″	Wood	1,050	50	950	1,060–910
SW3	40°04′53.352″	076°22′45.532″	Wood	2,030	40	1,990	2,100–1,890
SRTS-17	40°04′52.853″	076°22′45.609″	Wood	9,690	70	11,160	11,210–11,050 and 10,970–10,760

Char. Charcoal

Microfossil and Macrofossil Analyses

Bulk sediment and wood collected from major stratigraphic units were studied to reconstruct the floristic composition of the pre-historic floodplain area. Sub-samples of sediments were processed for plant macrofossils using the method of Lévesque et al. (1988). Samples for microfossil analysis were treated according to the methods of Riding and Kyffin-Hughes (2006). Treatments included hydrochloric acid (HCl), potassium hydroxide (KOH), and sodium hexametaphosphate ($Na(PO_3)_6$). Sediment samples were washed through a 125 μm sieve and retained on a 15 μm nitex screen. Material retained on the 125 μm was classified as macrofossil material. Microfossils retained on the 15 μm screen were stained with safranin and pollen aliquots were mounted in glycerol on glass slides. Pollen was identified with the use of published keys (Kapp 1969; Faegri and Iversen 1989).

We studied the buried wood by sectioning each sample with a razor blade along transverse, longitudinal, and radial planes. Wood that had dried was softened by soaking overnight in distilled water before sectioning. Wood shavings mounted in glycerol were studied using transmitted light microscopy to characterize the wood anatomy. Charcoal samples were mounted on stubs, gold-coated, and observed using a Zeiss DSM 962 digital scanning electron microscope. Wood and charcoal identification was carried out by reference to published works (Hoadley 1990) and the "InsideWood Database" (http://insidewood.lib.ncsu.edu/).

Results

Geology and Topography

Bedrock in the southern and eastern uplands of the Swarr Run Watershed include deformed metaquartzites of the EoCambrian Chickies Formation and phyllites, slates, and sandstones of the Cambrian Harpers/Antietam Formations. Underlying the topographically lower and flatter western and northern parts of the watershed are predominately carbonate units including the Cambrian Vintage, Kinzers, Ledger, and Zooks Corner Formations. The Vintage Formation is a thick bedded to massive, finely crystalline, gray dolomite. It commonly contains fine, wavy, siliceous, or argillaceous laminae. Some beds appear knotty or mottled (Kauffman 1999). The Kinzers Formation includes shale, limestone, and dolomite, while the overlying Ledger and Zooks Corner Formations are dominated by thin to massive-bedded, fine to coarsely crystalline dolomite. Vein quartz is common, especially in the clastic units. The field site, located in the northern and lowest part of the watershed is underlain by Vintage dolomites (Fig. 3.4).

The general gradient of the stream ranges from 3% in the headwaters of the creek where resistant units of the EoCambriam Chickies and Cambrian Antietam and Harpers Formations are exposed to between 0.2% and 0.3% along the lower reaches where the less resistant Vintage dolomites occur. The floodplain along the reach examined in this study is approximately 100 m wide and has a slope of 0.2%. The stream is incised into the floodplain sediments by up to 2 m everywhere along the study reach (Fig. 3.3). Bedrock is intermittently exposed in the channel, but mostly the channel floor consists of coarse grained gravel lag deposits. The channel includes straight and meandering sections strongly suggesting significant channel modification since European's settled the area. Examination of aerial photographs

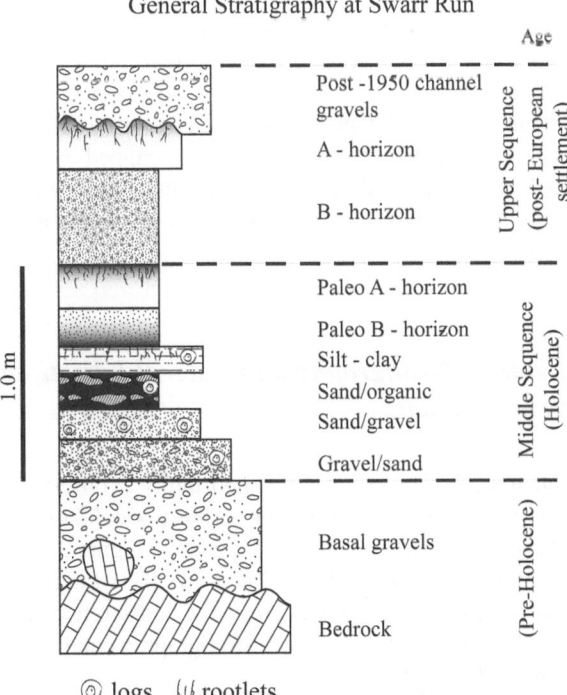

Fig. 3.4 Generalized stratigraphy at Swarr Run, Lancaster County, Pennsylvania

taken since 1936 shows that the straight reach extending for 300 m below the location of the trench (Fig. 3.1) was modified in the 1950s. The original channel was slightly longer and was located to the east of the present channel. This modification resulted in a change in the channel gradient from 0.17% to 0.19% producing a 10% increase in gradient. It is likely that other reaches of Swarr Run were similarly modified since European occupation in the early 1700s.

General Stratigraphy

The general floodplain stratigraphy is visible in numerous locations along the eroding banks of Swarr Run and illustrated in Fig. 3.4. We have divided the sedimentary sequence overlying the bedrock into basal gravels, middle sequence, upper sequence, and channel sediments. The bedrock EoCambrian and Cambrian formations are overlain by reddish coarse grained gravels. These 'basal' gravels range in thickness from 20 cm to several meters (and probably much thicker). A highly variable, approximately 1 m thick 'middle' sequence of sediments overlie the reddish gravels and extend up to a paleo-soil horizon. Typically, coarse grained gravels (lacking the reddish color) and sands up to 30 cm thick overlie the reddish gravels. In places a laterally continuous, but variably thick, organic-rich, fine-grained peaty layer lies directly on the reddish gravels. Woody fragments and logs up to 30 cm in diameter and many meters in length occur in these gravels, sands, and organic-rich layers above the reddish gravels. A laterally discontinuous, organic-rich silty layer overlies the peat horizon. This layer grades up into a silty, paleo B-horizon and is capped by a very distinct paleo A-horizon. Approximately 1–1.5 m of sediment, the 'upper' sequence overlies the paleo-soil horizon in every exposed floodplain sequence. Colluvial deposits (sediment transported by gravity) frequently inter-finger with these sediments. This sediment is capped by a 20–30 cm thick A-horizon. Up to 1 m of silt, sand, and gravel (channel sediments) occurs in isolated locations within the channel, but are especially abundant along the reach modified in the 1950s.

Detailed Stratigraphy and Lateral Variations Exposed in the Trench

The trench provides a unique opportunity to observe and document the vertical and especially the lateral, variation in the floodplain sequences, which are not readily observable in the streambank exposures. Figure 3.2 shows the details of the stratigraphy in the trench and Fig. 3.5 shows the stratigraphy, LOI, and grain size of the five sedimentary sequences measured at discrete points along the trench. The datum for the trench diagram and the stratigraphic columns is the present level of the base flow in the creek.

Bedrock Generally the Cambrian Vintage Formation is poorly exposed in the watershed, however, it was observed in the south end of the trench at a depth of 1.5 m below the surface (closest to the hillslope). The upper contact is sharp and dips gently towards the north and under the basal gravels of the floodplain sediments.

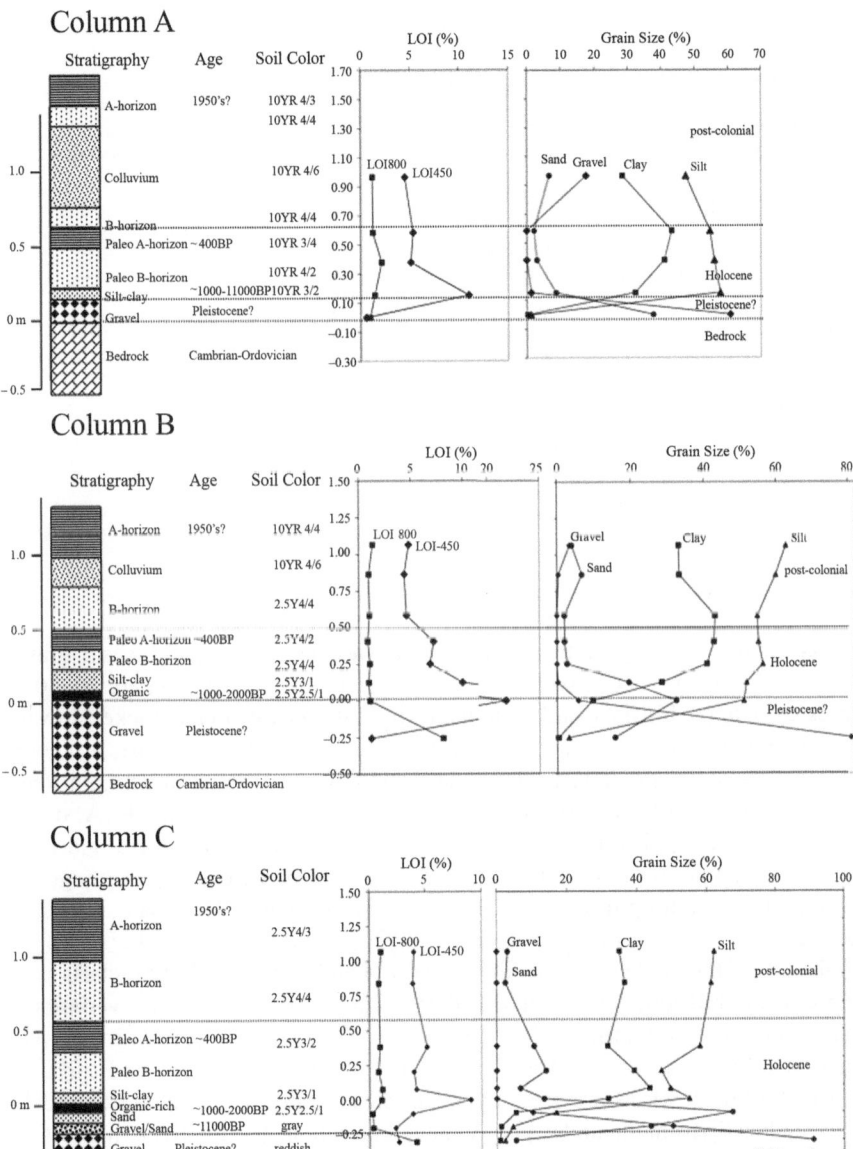

Fig. 3.5 Details of the stratigraphy, age, soil color, loss on ignition (LOI), and grain size of the sedimentary sequence at five locations in the trench (see Fig. 3.3)

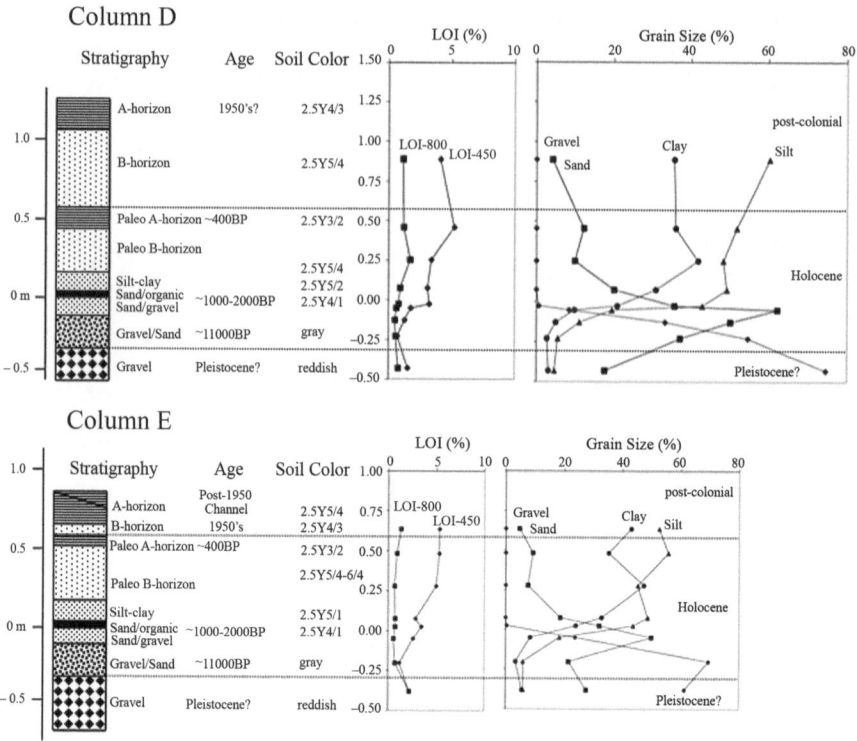

Fig. 3.5 (Continued)

Basal Gravels Overlying the bedrock is a variable thickness of basal gravels (Figs. 3.2 and 3.4). The gravels have distinctive reddish matrix and are 20–30 cm thick close to the hillslope and more than 1 m (but probably much thicker) towards the stream channel. This may be typical of the region. For example, at another location in Lancaster County, along the Santa Domingo Creek in Lititz, a trench revealed greater than 3 m of gravel without intersecting the bedrock (A. deWet, unpublished data). The basal gravels at Swarr Run are poorly sorted and stratified. The angular to sub-rounded clasts (sedimentary particles) range from gravels greater than 2 mm up to boulders 20–30 cm in diameter. The gravel-sized clasts greater than 2 mm comprise from 60% to 92% of these deposits (Fig. 3.4). The grain size of the gravels is generally between 4 and 8 cm. The largest clasts occur close to the hillslope, but otherwise there is no obvious systematic variation in the grain size either horizontally or vertically. The clasts are predominately vein quartz and metaquartzites, followed by finer grained argillaceous and carbonate clasts. Carbonate clasts are more abundant close to the slope and were probably derived locally from the underlying Vintage Formation. The relatively high LOI at 850°C of this deposit (2–6%) probably reflects the occurrence of limestone grains (Fig. 3.5). The reddish color of the basal gravels is the result of abundant iron oxide in the matrix. No woody material was found in these gravels. The upper contact of the basal gravel dips gently away from the hill-slope to the present stream channel. The contact is defined by a rapid change from the reddish iron stained gravels to a heterogeneous sequence of gravels, sands, and organic rich silts with no iron staining.

Middle Sequence Overlying the basal gravels is a laterally variable sequence of clastic sediments frequently containing abundant organic and woody material. Generally these sediments grade upwards from coarse gravels to sands, organic rich silts, and clayey silts. However, locally sands or organic-rich silts occur directly on the basal gravels. The upper boundary of this sequence is defined by a 10–20 cm thick paleo-soil horizon (Figs. 3.2 and 3.5).

Generally the middle sequence gravels are not as coarse as the underlying basal gravels. Gravel-sized clasts greater than 2 mm comprise from 50% to 70% of these deposits (Fig. 3.4). The grain size of the gravel ranges up to 6 cm. Weakly developed stratification is observed in the gravels. The gravels become finer upwards and include an increasing amount of wood fragments. None of the woody material is in growth position and it is rounded and abraded implying some transport. Thus the woody material appears to be rafted into these deposits. Lenses of fine-grained gravel and sand with abundant woody fragments and logs grade up into a variable thickness of dark gray organic rich, clayey-silts. This very distinct dark gray to black horizon has an organic matter content of between 8% and 22% and is interpreted as a paleo-organic soil horizon, a so-called 'peat'. This layer ranges from a few mm up to 35 cm in thickness and extends from the present stream channel to the break of the slope. Logs up to 30 cm in diameter and several meters in length are abundant in this horizon (Fig. 3.6a).

Organic rich silt/clay with rootlets drape over the buried peat horizon. This layer is thickest over the thickest part of the underlying peat, ranging up to 20 cm; perhaps resulting from sediment trapping around the vegetation. The organic content of these sediments decreases upward. The sediments do not appear to be finely layered, but extensive bioturbation (mixing of sediment particles due to faunal activity) may have destroyed any original layering. A few thin gravel lenses occur within the silts. A 15–20 cm, slightly organic-rich layer is evident across the entire sequence. It thins towards the stream channel and the slope. This layer also appears to be a buried organic soil horizon.

Upper Sequence A 40–60 cm thick layer of clayey silt with several thin, laterally discontinuous, layers of fine gravel is present above the upper organic rich paleosol. This sediment was probably deposited as a result of overbank flow given that it interfingers with colluvium derived from the slope. The uppermost part of the sequence is a 20–40 cm thick soil horizon with numerous active roots. The sediment is extensively bioturbated by worms (observed) and there is no evidence for fine scale laminations apart from the isolated thin gravel lenses.

Channel Sediments Up to 1 m of silt, sand, and gravel occurs in isolated locations within the channel, but these sediments are especially abundant along the reach mod-

Fig. 3.6 a Close-up of the transition from post-glacial gravels to organic-rich peat horizon and dark gray clay/silt deposits. Wood fragments up to several meters in length occur frequently within the organic rich layers. No trees in growth position were observed. **b** Lateral transition from floodplain sediments into the post-1950 channel deposits. The floodplain sediments were derived from overbank deposits and occasional slope derived flows. The sediment in the post-1950 channel includes fine-grained silts and sands and numerous gravel layers

ified in the 1950s (Fig. 3.4). Clasts are predominately vein quartz and metaquartzites, but numerous anthropogenic clasts such as glass, brick, ceramics, wood, and metal were observed. In several locations along the creek upstream of the trench, the channel sediments are comprised of coarse-grained gravels that appear to be point-bar deposits. Along the channel reach modified in the 1950s there are relatively more sands and silts with thin layers of fine-grained gravel compared to the point-bar deposits. The sediments are horizontally stacked and do not vary systematically across the channel (Fig. 3.6b). These deposits do not appear to be point-bar deposits, but rather, vertically accreting sediments in response to the modifications to the channel and watershed. The presently active channel is located in the basal gravels.

Chronology of Sediment Deposition

No datable material was found in the basal gravel deposits. A wood fragment from the coarse-grained gravels at the base of the middle sequence gave a calibrated radiocarbon age of 11,160 years BP (Figs. 3.2 and 3.4; Table 3.1). Two wood samples from the peat horizon in the middle sequence gave ages of 950 years BP (SW1) and 1,990 years BP (SW3) (Figs. 3.2 and 3.4). Sample SW1 was collected from a 15 cm diameter by 3 m long hardwood (possibly oak) log deposited horizontally (i.e., not in growth position) in the peat horizon with numerous other logs. SW3 was sampled from a smaller log located in the same horizon. A wood fragment located 10 cm below the dark gray paleosol horizon at the top of the middle sequence gave an age of 470 years BP. Charcoal collected 5 cm below the top of this dark gray paleosol gave an age of 450 years BP.

These results indicate that the basal gravels are likely Pleistocene or older. The middle sequence is Holocene in age and ranges around 12,000 years BP at the base to a minimum age of 300–400 years BP for the pre-European soil horizon at the top. The upper sequence was deposited after Europeans settled the area in the early 1700s. The channel sediments are therefore post-European settlement in age and are clearly post 1950s in the modified reach. The recent age of these sediments is evident by the abundance of recent artifacts including glass, brick, ceramics, and metal fragments.

Microfossil Analysis

Samples from the basal gravels were extracted for microfossil analysis, but were found to be barren of palynomorphs (pollen and spores). However, three sediment samples from the Holocene middle sequence contained measurable pollen and spores (Fig. 3.7). Two microfossil samples (SS41 and SS56) from the peat horizon had the highest percentage and diversity of palynomorphs. One sample (SS50) from the organic-rich silt layer that directly overlies the thick peat horizon also contained palynomorphs, but at much lower percentages. On average, *Carya* (hickory) and *Quercus* (oak) were the most abundant hardwood pollen types recovered from the peat, although a diverse array of other hardwood taxa such as *Acer* (maple), *Alnus* (alder), *Betula* (birch), *Fraxinus* (ash), and *Ulmus* (elm), albeit at low percentages were identified (Table 3.2). Pollen of the conifers *Pinus* (pine) and *Tsuga* (hemlock) were also recovered from the sediments although *Tsuga* pollen was recovered exclusively from the organic-rich silt. Pollen from herbaceous taxa such Cyperaceae (sedges), Compositae (sunflower family), Gramineae (grases) and Polypodiaceae (ferns) (e.g., *Dryopteris* [wood ferns] and *Pteridium* [bracken ferns]) were present in the peat horizon samples.

Macrofossil Analysis

We evaluated the sub-fossil wood preserved at various stratigraphic intervals exposed in the trench (Table 3.3). A mix of hardwood and conifer wood is present in

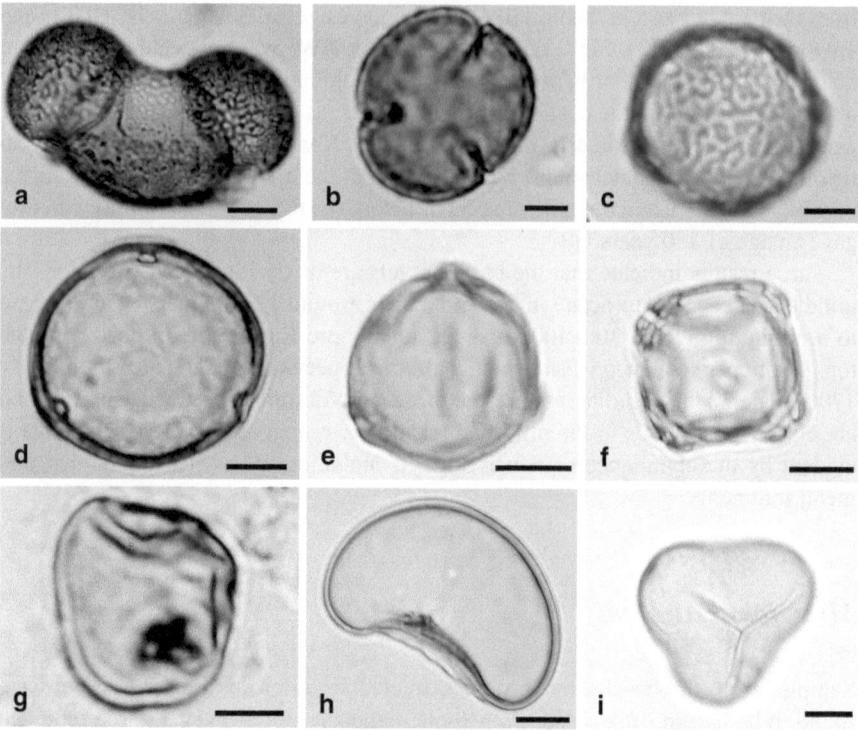

Fig. 3.7 A representative assemblage of palynomorphs identified from the late Holocene middle sequence sediments at Swarr Run, Lancaster County, Pennsylvania. **a** *Pinus* sp. (Haploxylon-pollen type), **b** *Tilia* sp., **c** *Ulmus* sp., **d** *Carya* sp., **e** *Betula* sp., **f** *Alnus* sp., **g** *Carex* sp., **h** *Pteridium* sp., **i** *Dryopteris* sp. Scale bar = 10 μm in all photos

Table 3.2 Pollen spectra from the late Holocene middle sequence sediments at Swarr Run, Lancaster County, Pennsylvania

Palynomorph type	Percentage of count
Pinus	30.8
Tsuga	2.3
Carya	17.5
Quercus	15.2
Juglans	4.2
Ulmus	1.5
Fraxinus	1.1
Tilia	1.1
Castanea	1.1
Betulaceae	4.2
Acer	1.9
Alnus	3.4
Compositae	3.4
Cyperaceae	2.7
Fern spores	7.6
Gramineae	1.7
Total palynomorph count	263

Table 3.3 Identification of buried wood recovered from trench at Swarr Run, Lancaster County, PA

Sample ID	Stratigraphic unit	Micro-anatomical traits	Identification and notes
TS-1, TS-3, TS-8	Middle sequence— upper sand/gravel	Solitary axial and tangential resin canals, earlywood to latewood transition gradual, cross-field pitting "windowlike", rays fusiform, ray tracheids with smooth walls	*Pinus* (white pine group)
TS-2	Middle sequence— upper sand/gravel	Resin canals absent, longitudinal parenchyma absent, earlywood to latewood transition abrubt, cross-field pitting piceoid/cupressoid, ray tracheids present	*Tsuga* sp.
TS-5	Middle sequence— lower gravel	Resin canals absent, longitudinal parenchyma absent, earlywood to latewood transition gradual to abrupt, cross-field pitting piceoid/cupressoid, narrow ray tracheids present	*Tsuga* sp.
TS-7	Middle sequence— lower gravel	Wood ring porous, rays conspicuous, tyloses abundant	*Quercus*?
Log-2	Middle sequence forest floor horizon	Wood ring porous, rays conspicuous, earlywood pores large, latewood pores small and indistinct, tyloses abundant, latewood vessels thin-walled	*Quercus alba*
TS-13	Middle sequence forest floor horizon	Wood ring porous, rays conspicuous, earlywood pores large, latewood pores small and indistinct, tyloses abundant, latewood vessels thin-walled	*Quercus alba*

the sediments at Swarr Run (Fig. 3.8). The stratigraphically lowest sample (TS5) comes from the boundary between the basal gravels and overlying fine grained sediments. This wood compares favorably to the wood of hemlock. Wood from the overlying Holocene channel gravels and overlying silts were a conifer and hardwoods mix. Three samples (TS1, TS3, and TS8) were identified as conifer wood belonging to the genus *Pinus* and are most likely *Pinus strobus* (eastern white pine). One sample (TS2) was comparable to the wood of *Tsuga*. Wood samples (TS11, TS12, and TS13) from the peat horizon were exclusively those of hardwoods. Radiocarbon dates from one hardwood (SW1) indicate that it dates from 950 years BP. This sample and TS13 have wood anatomy comparable to *Quercus*, and in particular *Quercus alba* (white oak). Other samples from this layer were not well enough preserved to identify to a specific taxon. One charcoal sample recovered from sediments associated with TS12 was also identified as a hardwood.

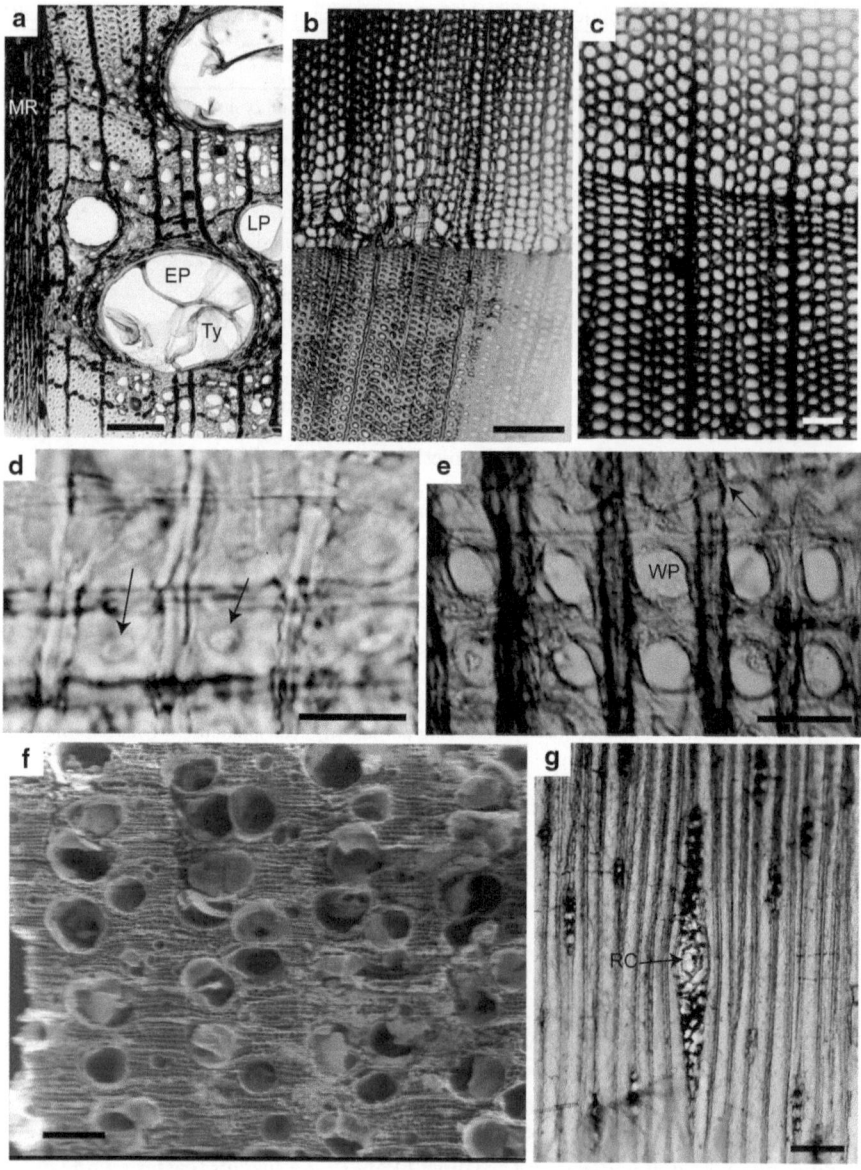

Fig. 3.8 Representative wood anatomy of buried wood and charcoal from the late Holocene middle sequence sediments at Swarr Run, Lancaster County, Pennsylvania. **a** Transverse section of sample TS-13 (*cf. Quercus alba* [white oak]). Note large earlywood pore (*EP*) with tyloses (*Ty*), small latewood pores (*LP*) and multiseriate rays (*MR*); scale bar = 100 μm. **b** Transverse section of sample TS-5 (*cf. Tsuga*) across a growth ring boundary. Earlywood above and latewood below; scale bar = 200 μm. **c** Transverse section of sample TS-3 (*cf. Pinus*) across a growth ring boundary, earlywood above and latewood below; scale bar = 100 μm. **d** Radial section of sample TS-5 indicating cupressoid-type cross-field pits with distinct boundary (*arrows*); scale bar = 25 μm. **e** Radial section of sample TS-3 indicating window-like cross-field pitting (*WP*) and ray tracheid boundary (*arrow*); scale bar = 25 μm. **f** Transverse section of charcoal recovered from sediments associated with wood sample TS-12; scale bar = 150 μm. **g** Tangential section of sample TS-3 indicating fusiform ray with central resin canal and thin-walled epithelial cells; scale bar = 100 μm

Discussion

The fluvial sedimentary record of the section of Swarr Run that we studied provides a greater than 10,000-year long archive of terrestrial environmental change in an upland catchment. Although limited in terms of the areal extent and subject to biases inherent in the study of fluvial deposits; these sediments nonetheless offer an opportunity to study the complex and dynamic evolution of a mid-Atlantic floodplain environment, especially the relationship between the flora and sediment supply through time. To first order, this provides a vegetation palette for restoration. It is clear that changes in the fluvial system have resulted from both natural and anthropogenic drivers of change. Some sedimentary changes track secular changes in the paleoenvironment and the paleoclimate of the region. Other changes are a consequence of anthropogenic activity in the watershed. Regardless of the driver, the consequences and magnitude of such change and the implications for future restoration and management are archived in the sedimentary record, which we explore below.

Pre-colonial Depositional Environment and Sedimentation Rates

The particle size and texture of the sediments we studied preserved fluvial and alluvial activity representative of at least three or more phases of deposition punctuated by periods of erosion. The coarse-grained basal gravels are remarkably homogeneous and persistent. In fact, these gravels appear at the base of floodplain sediments and were observed in numerous locations in Lancaster County (A. deWet, unpublished data). The shape, size, and sorting of these gravels indicates that they are fairly proximal and were derived either from hillslope processes and the erosion of metaquartzites in the headwaters or from the slow dissolution and weathering of quartz veins in nearby carbonate strata. The increase in the abundance of carbonate clasts closer to the hillslope indicates that some of the clasts were derived locally from the surrounding Vintage Formation limestone and are extremely proximal. Away from the slope, the increasing abundance of vein quartz and metaquartzites implies increased transportation and reworking of these materials by flowing water and subsequent deposition as basal gravel lag deposits.

Despite the widespread occurrence of the basal gravels, no formal name has been assigned to them. Outwash gravels that occur as isolated terrace deposits have been described along the major rivers including the Susquehanna (Pazzaglia and Gardner 1993); however, fluvial deposits, especially basal gravels in the valley bottoms of streams in the non-glaciated areas have not been well studied (Crowl and Sevon 1999). Recent studies have focused on the overlying pre-colonial and post-colonial sediments, for example Walter and Merritts (2008). While it is not possible to directly correlate the fluvial gravels with the terrace gravels, age data suggest that the basal gravels in the small streams might correlate with the Pleistocene Qt1-Qt5 terrace gravels of Pazzaglia and Gardner (1993) preserved along the Susquehanna River.

Because we were unable to determine the exact age range of the reddish basal gravels it was not possible to determine a deposition rate. It seems probable that long-term climate-driven erosion and in-place weathering (Costa and Cleaves 1984; Stanford et al. 2001; Dixon et al. 2009), as well as low-frequency, but high-magnitude events such as floods (Eaton et al. 2003), played a role in the origin of the basal gravels within the catchment. Given the geologically slow weathering rate of vein quartz it is probable that these gravels represent at least tens of thousands of years of deposition and reworking on the landscape. They are in essence the remnants of a long-term Piedmont landscape denudation process (Stanford et al. 2001).

The middle sequence of sediments represents the deposition of a relatively stable Holocene floodplain surface, probably typical of the streamside areas of Swarr Run prior to European settlement of the watershed. As the last ice age waned, forests developed on the periglacial sediments, and soon dominated the landscape (Watts 1979; Davis 1983; Abrams 2002). Invariably, swampy areas developed adjacent to stream channels in the valley, forming a bottomland forest. Peat, organic-rich clays, and numerous logs were deposited in these low-energy environments (Fig. 3.6a).

Our analyses of the plant micro- and macrofossils provide some, albeit limited, knowledge of the composition of the flora in the watershed and on the floodplain. There is a lack of detailed information on the floristic composition of Piedmont region of Pennsylvania throughout the Holocene to which we can compare our results. However, general ideas about the composition of the dominant vegetation have been assembled (Williams et al. 2004) and these correspond to the composition of the pollen and macrofossils recovered at Swarr Run. For example, by 9,000 years BP, oak-dominated woodlands, a component of the extensive oak-hickory-chestnut forest assemblage (Braun 1950; Dyer 2006) are present in the lowland Piedmont and persist throughout the Holocene (Abrams 2002). Our findings that oak, hickory, and pine dominate the arboreal (tree) pollen spectra of Swarr Run's buried peat horizon support this notion. Notably, our recovery of buried white oak wood in the organic-rich deposits of the middle sequence provides the first macrofossil evidence that white oak was a component of the late Holocene floodplain forests. Interestingly, chestnut is not a principle component of the pollen sampled in this layer. This is not altogether unexpected given that chestnut is rarely found in wet locations or on limestone soils (Black and Abrams 2001).

The general floristic composition reconstructed from the palynomorphs and buried wood recovered from the buried peat horizon in the middle sequence aligns fairly well with estimates of forest composition of the late Holocene, pre-European settlement period. For example, Black and Abrams (2001) utilized historic land survey records containing witness tree data to reconstruct the species composition and distribution of forest types in Lancaster County, Pennsylvania. For areas of lowland Piedmont in which Swarr Run is located, hickory and oak were most abundant prior to and coincident with extensive clearing for agriculture. Pines and chestnut were only abundant in Piedmont upland areas. The existence of pine pollen and wood in the Swarr Run Middle Sequence deposits suggests that pines may have occupied some lowland areas in the Swarr Run catchment. However, it seems more likely that the pine wood is transported. Occasionally over-bank or channel sands and gravels are also preserved in the middle sequence. Indeed, the pine wood was recovered from such gravels suggesting that this material was probably transported into the floodplain during periods of higher water flows.

The herbaceous and shrub pollen recovered from the buried peat horizon at Swarr Run includes sedge and alder pollen as well as monocot and dicot angiosperm pollen. Fern spores were also present. It is likely that the hydric soils in the riparian areas without extensive tree cover supported the growth of a wide variety of herbaceous plants and woody shrubs. Extensive studies of macrofossils, such as wood and seeds, from other late Holocene buried floodplains in the region are lacking, although preliminary results documenting abundant sedge seeds recovered from a nearby late Holocene buried organic soils support this idea (Davies et al. 2006; Voli et al. 2009).

In places, organic-rich, gray clays were deposited on top of the peat horizon. It is possible that this represents an increase in sediment deposition prior to European colonization in the area. This might be related to climate changes, the influence of the Native Americans, or may represent lateral variations in the depositional environment in the valley. These sediments are capped with a laterally extensive 20–30 cm thick paleosol horizon that dates from just before the colonial period.

The 11,160 year BP age from a log in the Holocene gravels and the 1,000–2,000 year ages for the overlying peat horizon implies a very slow deposition rate for the Holocene gravels of less than 0.03 cm per year (Fig. 3.9). These gravels represent the equilibrium base level of the stream channel and were probably reworked within the channel of the stream over thousands of years. The paleo-A horizon at the top of the middle sequence, dated at around 400 years BP, thus represents the pre-European settlement surface. The sedimentation rates of the finer-grained sediments and organic-rich layers prior to European settlement are estimated to be between 0.03 and 0.1 cm per year (Fig. 3.9). The Holocene floodplain surface, characterized by slow accumulation rates (0.03 cm per year) is a feature that is found at other locations in Lancaster County (deWet, unpublished data; Walter and Merritts 2008).

There is no evidence in the excavated trench of the late Holocene stream channel itself. Prior to the 1950s the channel was located to the east of the present channel and likely implies that the late Holocene channel was also located to the east of the trench and the paleosol represents the top of the adjacent floodplain. Variability in the earlier Holocene sediments exposed in the trench implies either a change in the overall depositional environment or local lateral migration of the stream system. The stream channel was probably never more than 1 m below the adjacent floodplain and both were hydraulically connected. The rapid accumulation of sediment during post-European settlement time probably resulted in an increasingly disconnected floodplain.

Various authors have suggested that Native Americans influenced the landscape by forest clearance and farming (Gooding 1971; Alexander and Prior 1971; Abrams and Nowacki 2008). Fire was also a component of the landscape change as is evidence by abundant charcoal in the paleo A-horizon (dated at 450 years BP). It is unclear whether forest clearance and farming by Native Americans or climate change

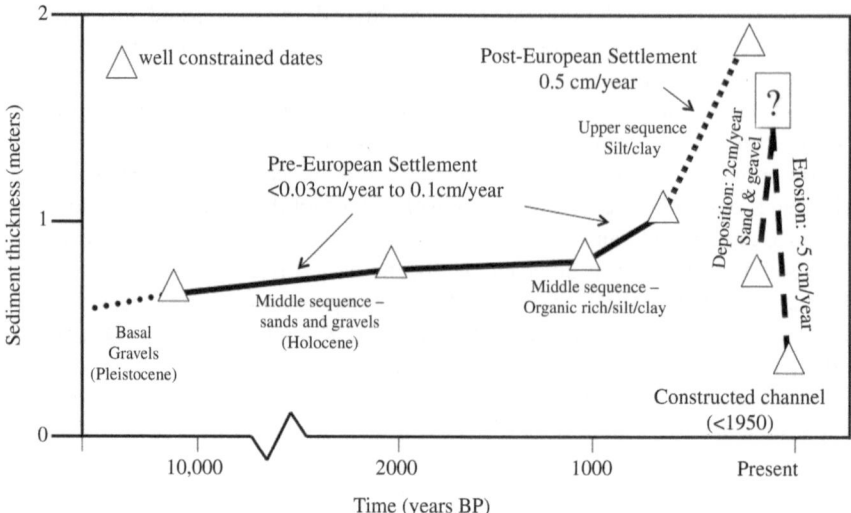

Fig. 3.9 Sediment deposition and erosion rates based on the age and thickness of the sediments preserved in the study reach floodplain. The deposition rate of the basal gravels could not be determined. The pre-European settlement deposition rate varied between less than 0.03 and 0.1 cm per year. The post-European deposition rate was up to 0.5 cm per year. Deposition rates of the channel sediments were probably up to 2 cm per year immediately after the excavation of the new channel but eventually transitioned into rapid erosion averaging up to 5 cm per year until the channel intersected the coarse-grained Pleistocene gravels. Immediately prior to the restoration project in 2003 the vertical channel erosion was probably negligible with most erosion occurring laterally (*bank erosion*) (see Fig. 3.11a). The restoration project has reduced the bank erosion over the last few years, at least along the restored reach (see Fig. 3.11b)

was the principle factor influencing the deposition of sediment during the pre-European settlement Holocene. In any case, the deposition rate of 0.03–0.1 cm per year was significantly less than the post-European settlement rate of 0.5 cm per year.

Post-European Settlement Landscape Evolution, Channel Modifications, and Sedimentation Rates

European colonization of the area began in the early 1700s—Lancaster City was founded in 1729 (Mombert 1869). The landscape changed rapidly due to deforestation and rapidly expanding agricultural landuse. By the 1750s land clearance and settlement of the lowland areas was largely complete (Lord 1975; Black 1998). Land clearing, combined with poor farming practices, resulted in increased soil erosion and sediment load to the streams, significantly modifying the stream channels and adjacent floodplains. As recently as the 1940s deep erosional gullies in agricultural fields were common in Lancaster County. By the late 1940s the valleys had accumulated over one meter of post-European settlement sediment (deWet and Tomlinson 2003). Sediment deposition in the floodplain included overbank clays and occasional sands and gravels, as well as slope-derived deposits. Only in the 1960s and 1970s did new farming practices such as strip cropping, contour plowing, and terracing start to reduce the sediment input from farms into local streams. Because the modern stream channel is deeply incised into the floodplain sediments and flooding seldom results in overbank flows, it is unlikely that significant sediment was deposited on the floodplain since the modification of the channel in 1950. The average deposition rate of the post-European settlement sediments is approximately 0.5 cm per year (100 cm per 200 years), or at least a five times increase in the pre-European settlement sediment deposition rate. This estimate is based on one location, but is consistent with estimates from other locations in Lancaster County and across the Piedmont (Jacobson and Coleman 1986).

Soon after 1950, when 200 m of Swarr Run in the study area was channelized and straightened, nearly one meter (vertical thickness) of channel deposits had accumulated in the new channel (Figs. 3.1 and 3.6b). This implies that initially after the new channel was created more than 2 cm of sediment was deposited per year (Fig. 3.9). By the 1980s the stream began to incise into these very recent channel sediments as well as continuing to erode its banks. Erosion rates may have reached up to 5 cm per year. The initial rapid deposition rate reflects the large amount of readily available sediment resulting from the extensive re-engineering of the stream channel upstream. This sediment was reworked downstream and resulted in very high deposition rates in the channel. The subsequent switch to an erosional regime might in part have resulted from the depletion of this available sediment, but may also reflect changes in the watershed.

GIS analysis of the landuse changes since the 1930s indicate that extensive landuse change occurred especially between 1960 and 1990 (Fig. 3.10). The transition from largely agricultural to residential, industrial, and commercial landuse increased the percentage impervious cover, which may have increased storm water runoff and resulted in larger and more frequent floods. This phenomenon, documented in urbanizing watersheds in the mid-Atlantic region (Beighley and Moglen 2002; Palmer et al. 2002; Colosimo and Wilcock 2007), would have initially supplied large amounts of water and sediment to the stream system. Later, as the landuse change slowed and sediment controls were introduced, sediment deposition probably declined, but surface run-off remained elevated or increased (Wolman and Schick 1967; Allmendinger et al. 2007). Rapid channel incision ensued, with high bed shear stress cutting down to the paleo-channel gravels. Prior to the stream restoration in 2003, the stream was deeply incised and was no longer connected to the floodplain. Since the channel gravels prevented further incision, the stream was widening its channel through bank erosion producing a wide, shallow channel typical of other streams in urbanizing watersheds (Hession et al. 2003). This phase represents the establishment of a quasi-equilibrium state that Leopold (1973) hypothesized would develop in channels responding to new flow regimes. Locally, channel widening due to bank erosion was exacerbated by the destabilizing effect of grazing animals along the channel.

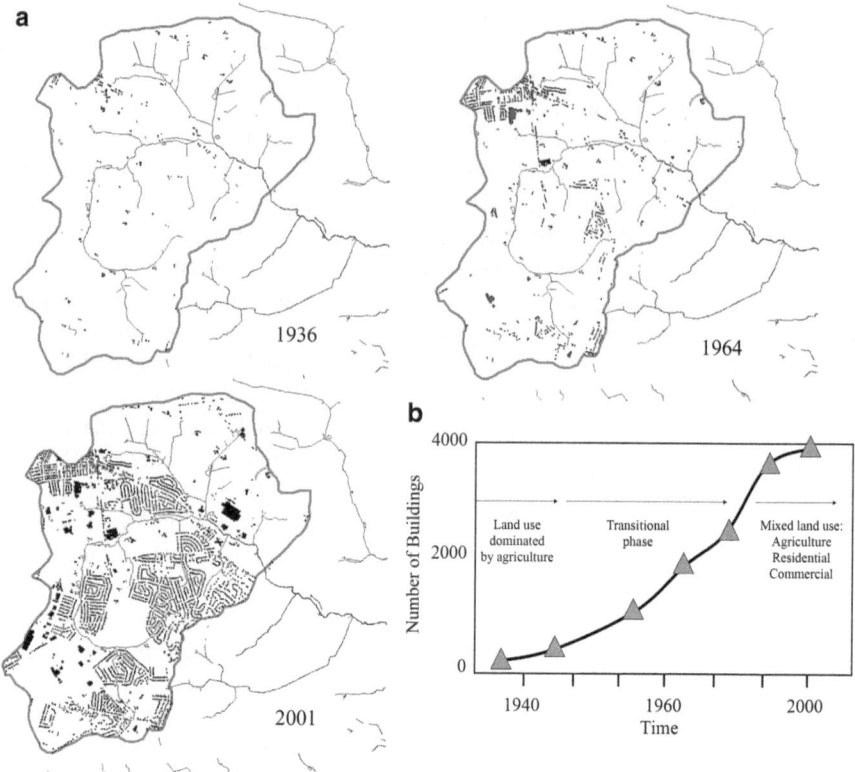

Fig. 3.10 a GIS maps showing distribution of buildings in the Swarr Run Watershed in 1936, 1964, and 2001. The buildings were derived from aerial photography (Tomlinson 2003). There was a dramatic increase in development in the 1970s. This change in the landscape probably resulted in an overall decrease in the sediment input into the stream channels and an increase in water discharge. This would explain the initial deposition of sediment in the post-1950 channel and the subsequent transition into net erosion and incision. **b** *Buildings* versus *Time* showing the rapid development between 1960 and 1990

Recent Restoration Efforts

By the late 1990s the studied reach of Swarr Run was arguably one of the most heavily impacted streams in Lancaster County. However, during this time the local population became increasingly aware of the importance of healthy streams, and a local watershed alliance formed to organize and fund stream restoration. During August and September 2003, a 1.1 km (0.7 mile) reach of Swarr Run located downstream of the State Street bridge, underwent restoration by the Lancaster County Conservation District under the supervision of Matt Kofroth (Figs. 3.1 and 3.11). Funding was obtained through the Pennsylvania Growing Greener program.

The restoration project involved reshaping and modifying the streambanks to reduce erosion. Streambank angles were lowered and banks were stabilized to reduce

Fig. 3.11 a View downstream (south-east) of the 1950 channelized reach of Swarr Run before restoration efforts in 2003. Notice the location of the trench. **b** Post-restoration view of the same area

bank erosion. Bank stabilization involved a variety of techniques including logs with rootwads, boulders, geofabrics, plantings, and rip-rap in high erosion areas. The channel sediments that had accumulated since 1950 and part of the adjacent floodplain, were partially excavated to reduce their elevation in order to create a new floodplain that is hydraulically reconnected to the stream channel. A fenced riparian buffer was established to keep the grazing animals off the banks and out of the stream channel. With time, the sediment record in this part of Swarr Run may again reflect watershed processes, rather than reflecting local agricultural practices.

Broader Implications

The sediments and fossil plant materials we analyzed in this study reveal that the present day riparian environment differs greatly from that of the pre-European settlement period. It has been known for some time that near-stream environments such as floodplains play a critical role in the ground- and steam-water nutrient dynamics (Karr and Schlosser 1978), and the changes we document may be important for water-quality management; particularly in this watershed (Peterjohn and Correll

1984; Burt and Pinay 2005). For example, denitrification that occurs in riparian soils prevents upland-derived nitrate from reaching stream waters. This process occurs in anoxic riparian soils and can be effective for nitrogen load reduction in restored floodplain wetlands and forests (Muenz et al. 2006). Nevertheless, the effectiveness of riparian areas in reducing nitrogen input to stream water may be inconsistent and depend on site-specific factors such as soil type, hydrology, and even carbon supply in the soil (Mayer et al. 2007; Gurwick et al. 2008). To this end, detailed studies such as ours, on the sub-surface properties of the riparian environment may set the stage for promoting enhanced nutrient reduction capacity at a specific site.

Sediment yield and the associated nutrient flux are also major drivers of stream impairment and a unique aspect of the post-European settlement floodplain topography. The present day floodplain at Swarr Run sheds sediment and nutrients into the active stream channel at rates that probably exceed that of the pre-European settlement system. Floodplain deposition, a predominating process on pre-European settlement floodplain surfaces is basically non-existent on the present day floodplain that is perched nearly a meter above the present stream base-level. Erosion of unstable streambanks and the migration of the post-1950 channel gravels are the two fluvial processes that dominate the present system. Fluvial sediment properties can modify the forms and amounts of phosphorous in steam water (McDowell et al. 2001). Considering that the particle size distribution of the unstable and eroding post-European settlement bank material is significantly different than the pre-European settlement floodplains it is likely that the within stream chemical transformations of nutrients may be different today than in the past. This notion would seem to be borne out by watershed studies that document landuse (crop land versus forest) as a strong correlate of within stream sediment phosphorous release (McDowell et al. 2003).

The shift in the dominant vegetation type in the Swarr Run Watershed is a factor that has very real implications for not only nutrient management, but also riparian restoration efforts. The paleoecological information stored in the sediments at Swarr Run provides a record of the vegetation composition of the Swarr Run valley prior to European settlement. Abundant buried wood, suggests a mixed forest of hardwoods and conifers in the valley bottom. The size of the buried logs suggests that at least some trees were growing close to the active channel. Relatively well developed organic-rich, hydric soils and pollen evidence for shade intolerant, obligate wetland plants such as *Carex* suggest some areas of the floodplain may have been relatively open on the wettest sites. Although the pace of change in the dominant vegetation awaits a more robust age-depth model for the sediments that we analyzed, our results suggested that more detailed studies of the paleoflora of the Holocene sediments would be feasible. With more detailed analyses, the archive of vegetation stored in the Swarr Run sediments may provide an appealing restoration benchmark for present-day degraded systems.

However, several factors present significant challenges to the long-term re-establishment of pre-European settlement vegetation assemblage in riparian areas. Manipulation of the hydrologic regime via the construction of dams, channelization and channel diversion, and irrigation has served to decouple riparian areas from water sources. Moreover, the long-term legacies of interactions between hydrologic regime modifications, local landuse history such as long-term grazing, and regional environmental change on riparian forests are generally poorly quantified, but likely to have a significant influence on the success of a restoration effort (Bedford 1999). In the case of Swarr Run, human alteration of fire regimes in the valley bottom, the existence of copious exotic and invasive plant species, and the supply of excess nutrients and soil changes associated with long-term agricultural activity all present restoration challenges beyond simple restoration of site hydrology (Zedler 2000, 2003).

Conclusions

The sedimentary sequences preserved in streams and floodplains provide detailed insights into the spatial and temporal variations in the dynamic processes affecting the associated watershed. Temporal scales can vary from millions of years to a few years or decades, while spatial scales can vary from the whole watershed down to individual stream reaches. In Lancaster County, slow denudation of the land surface since the last period of mountain building has removed many kilometers of material (Stanford et al. 2001). In Swarr Run the thick basal gravels reflect the last phase of this long-term process. Dissolution of the EoCambrian and Cambrian carbonates and erosion of the associated clastic units have yielded vein quartz and lithic fragments that accumulated as basal gravels over one or more glacial cycles. Since the last ice age a complex sequence of sediment has accumulated in the floodplain. Fluvial gravels and sands dating from the early Holocene are overlain by finer-grained floodplain sediments and adjacent organic rich wetland deposits. Earlier cycles of floodplain sequences may have accumulated during previous interglacial periods only to be reworked and destroyed by later erosional episodes. The Holocene sediments (middle sequence) record the evolution of the depositional environment from an initial periglacial (freezing) environment to a more stable forested watershed with a stream channel surrounded by a wet hardwood forest and slow accumulation of sediment in the adjacent floodplain. This sequence is capped with a paleo-A horizon, which dates to approximately 400 years BP. This horizon represents the land surface immediately prior to European settlement. The overlying upper sequence records a substantial increase in the rate of sediment deposition during the post-European settlement period resulting from widespread changes in the land-cover and hydrological characteristics of the watershed. Initially deforestation and agriculture profoundly altered the watershed and more recently suburban and commercial development altered the sediment and water supplies. These processes operated over time scales of decades to centuries. Superimposed on these watershed-scale processes were spatially limited, but equally important processes such as channelization and local farming practices.

Detailed analysis of floodplain sediment characteristics allows for an understanding of the watershed processes operating through time. These are dynamic systems, constantly changing in response to multiple interacting processes operating on various spatial and temporal scales. Clearly many processes, such as climate change, operated over a wide geographic area and thus are common to other watersheds in the region, while other processes, such as channelization, may be unique to a particular watershed or even to a specific reach of a stream. Without this understanding, our ability to predict how riparian systems, especially restored ones, will respond to these processes will be difficult, if not impossible. Deciding on restoration strategies and accurately predicting the outcome of restoration efforts is likely to be difficult without knowledge of the historical processes operating on a stream. Compounding this challenge is the likelihood that factors such as landuse and climate will continue to change in the future.

Acknowledgements We wish to thank the faculty and students involved in the Keck Geology Consortium—Little Conestoga Watershed project: Dorothy Merritts, Jeffrey Marshall, Steve Weaver, Garrett Bayrd, Abby Bowers, Aaron Davis, Jennifer Fallon, Nancy Harris, Ashley Hawes, Kyle Cavanaugh, and Lauren Manion. Matt Kofroth of the Lancaster County Conservation District provided access to the site, Flyway Excavating Company excavated the trench, and Christopher Sommerfield at the Graduate College of Marine Studies, University of Delaware provided access to the sedigraph.

References

Abrams MC (2002) The postglacial history of oak forests in eastern North America. In: McShea WJ, Healy WM (eds) The ecology and management of oaks for wildlife. John Hopkins University Press, Baltimore, pp 34–45

Abrams MD, Nowacki GJ (2008) Native Americans as active and passive promoters of mast and fruit trees in the eastern USA. Holocene 18:1123–1137

Alexander CS, Prior JC (1971) Holocene sedimentation rates in overbank deposits in the Black Bottom of the lower Ohio River, southern Illinois. Am J Sci 270:361–372

Allan JD (2004) Landscapes and riverscapes: the influence of land use on stream ecosystems. Annu Rev Ecol Evol Syst 35:257–284

Allmendinger NE, Pizzuto JE, Moglen GE, Lewicki M (2007) A sediment budget for an urbanizing watershed, 1951–1996, Montgomery County, Maryland, USA. J Am Water Res Assoc 43:1483–1498

Bayrd GB, Davis AG, Harris NJ (2002) A geomorphic interpretation of the Swarr Run Watershed of Lancaster Pennsylvania. In: Proceedings of the 15th Annual Undergraduate Research Symposium, Keck Geology Consortium, pp 170–173

Bedford BL (1999) Cumulative effects on wetland landscapes: links to wetland restoration in the United States and southern Canada. Wetlands 19:775–788

Beighley RE, Moglen GE (2002) Assessment of stationarity in rainfall-runoff behavior in urbanizing watersheds. J Hydrolog Eng, ASCE 7:27–34

Black BA (1998) Physiographic analysis of witness tree distribution and surveyor bias in the pre-European settlement forest of Lancaster county, PA. M.S. thesis, The Pennsylvania State University, University Park, Pennsylvania

Black BA, Abrams MD (2001) Analysis of temporal variation and species-site relationships of witness tree data southeastern Pennsylvania. Can J For Res 31:419–429

Braun EL (1950) Deciduous forests of eastern North America. The Blackburn Press, Caldwell, p 596

Burt TP, Pinay G (2005) Linking hydrology and biogeochemistry in complex landscapes. Prog Phys Geogr 29:297–316

Colosimo MF, Wilcock PR (2007) Alluvial sedimentation and erosion in an urbanizing watershed, Gwynns Falls, Maryland. J Am Water Res Assoc 43:499–521

Costa JE, Cleaves ET (1984) The Piedmont landscape of Maryland: a new look at an old problem. Earth Surf Processes Landf 9:59–74

Crowl GH, Sevon WD (1999) Quaternary, vol 1. Pennsylvania Geologic Survey Special Publication, pp 224–231

Davis MB (1983) Holocene vegetational history of the eastern United States. In: Wright HE Jr (ed) Late quaternary environments of the United States, vol 2. The University of Minnesota Press, St. Paul, pp 166–181

Davies GM, deWet AP, Williams CJ, Wilson MD (2006) Dillerville swamp: a major wetland in SE Pennsylvania? Northeastern section–41st Annual Meeting, vol 38. Geological Society of America Abstracts with Programs, p 88

deWet AP, Tomlinson J (2003) Stream processes, land-use change and sediment supply, Lancaster County, PA, vol 34. GSA Abstracts with Programs, p 314

Dixon JL, Heimsath AM, Kaste J, Amundson R (2009) Climate-driven processes of hillslope weathering. Geology 37:975–978

Dyer JM (2006) Revisiting the deciduous forests of eastern North America. BioScience 56:341–352

Eaton LS, Morgan BA, Kochel RC, Howard AD (2003) Role of debris flows in long-term landscape denudation in the central Appalachians of Virginia. Geology 31:339–342

Faegri K, Iversen J (1989) Textbook of pollen analysis, 4th edn. Wiley, New York, p 328

Gellis AG, Banks SL, Langland MJ, Martucci SK (2004) Summary of suspended-sediment data for streams draining the Chesapeake Bay Watershed, water years 1952–2002. U.S. Geological Survey Scientific Investigations Report 2004–5056, pp 1–59

Gooding AM (1971) Postglacial alluvial history in the Upper White-Water basin, southeastern Indiana, and possible regional relationships. Am J Sci 271:389–401

Gurwick NP, Groffman PM, Yavitt JB, Gold AJ, Blazejewski G, Stolt M (2008) Microbially-available carbon in buried riparian soils in a glaciated landscape. Soil Biol Biochem 40:85–96

Hassett B, Palmer MA, Bernhardt ES, Smith S, Carr J, Hart DD (2005) Restoring watersheds project by project: trends in Chesapeake Bay tributary restoration. Front Ecol Environ 3:259–267

Hession WC, Pizzuto JE, Johnson TE, Horowitz RJ (2003) Influence of bank vegetation on channel morphology in rural and urban watersheds. Geology 31:147–150

Hoadley BR (1990) Identifying wood: accurate results with simple tools. The Taunton Press, Newtown, p 240

Jacobson RB, Coleman DJ (1986) Stratigraphy and recent evolution of Maryland piedmont flood plains. Am J Sci 286:617–637

Johnstone E, Macklin MG, Lewin J (2006) The development and application of a database of radiocarbon-dated Holocene fluvial deposits in Great Britain. Cateña 66:14–23

Kapp RO (1969) How to know pollen and spores. W.C. Brown, Dubuque, p 249

Karr JR, Schlosser IJ (1978) Water resources and the land-water interface. Science 201:229–234

Kauffman ME (1999) Eocambrian, Cambrian, and transition to Ordovician. In: Shultz CH (ed) The geology of Pennsylvania. Special Publication of the Geological Survey of Pennsylvania. Harrisburg, Pensylvania, pp 59–73

Langland MJ, Hainly RA (1997) Changes in bottom-surface elevations in three reservoirs on the lower Susquehanna River, Pennsylvania and Maryland, following the January 1996 flood—implications for nutrient and sediment loads to Chesapeake Bay. United States Geological Survey Water-Resources Investigations Report 97-4138, pp 1–39

Langland MJ, Edwards RE, Sprague LA, Yochum S (2001) Summary of trends and status analysis for flow, nutrients, and sediments at selected nontidal sites, Chesapeake Bay basin, 1985–1999. United States Geological Survey Open-File Report 01-73, pp 1–49

Leopold LB (1973) River channel change with time: an example. Geol Soc Am Bull 84:1845–1860

Leopold LB, Huppman R, Miller A (2005) Geomorphic effects of urbanization in forty-one years of observation. Proc Am Philos Soc 149:349–371

Lévesque PEM, Dinel H, Larouche A (1988) Guide to the identification of plant macrofossils in Canadian peatlands. Land Resource Centre, Ottawa, Ontario. Research Branch, Agriculture Canada. Publication No. 1817, p 65

Loper CA, Davis RC (1998) A snapshot evaluation of stream environmental quality in the Little Conestoga Creek Basin, Lancaster County, Pennsylvania. United States Geological Survey Water Resources Report 98-4173, pp 1–8

Lord AC (1975) The pre-revolutionary agriculture of Lancaster County, Pennsylvania. J Lancaster Cty Hist Soc 79:23–42

Macklin MG, Lewin J (2003) River sediments, great floods and centennial-scale Holocene climate change. J Quat Sci 18:101–105

Macklin M, Benito G, Gregory K, Johnstone E, Lewin J, Michczynska D, Soja R, Starkel L, Thorndycraft VR (2006) Past hydrological events reflected in the Holocene fluvial record of Europe. Cateña 66:145–154

Mayer PM, Reynolds SK, McCutchen MD, Canfield TJ (2007) Meta-analysis of nitrogen removal in riparian buffers. J Environ Qual 36:1172–1180

McDowell RW, Sharply A, Folmar G (2001) Phosphorus export from an agricultural watershed: linking source and transport mechanisms. J Environ Qual 30:1587–1595

McDowell RW, Sharply AN, Folmar G (2003) Modification of phosphorus export from an eastern USA catchment by fluvial sediment and phosphorus inputs. Agric Ecosyst Environ 99:187–199

Mombert JI (1869) Authentic history of Lancaster County in the State of Pennsylvania. J.E. Barr, Lancaster, p 792

Muenz TK, Golladay SW, Vellidis G, Smith LL (2006) Stream buffer effectiveness in an agriculturally influenced area, southwestern Georgia: responses of water quality, macroinvertebrates, and amphibians. J Environ Qual 35:1924–1938

Naiman RJ, Bunn SE, Nilsson C, Petts GE, Pinay G, Thompson LC (2002) Legitimizing fluvial ecosystems as users of water: an overview. Environ Manag 30:455–467

Palmer MA, Moglen GE, Bockstael NE, Brooke S, Pizzuto JE, Wiegand C, van Ness K (2002) The ecological consequences of changing land use for running waters: the suburban Maryland case. Yale Environ Sci Bull 107:85–113

Paul MJ, Meyer JL (2001) Streams in the urban landscape. Annu Rev Ecol Syst 32:333–365

Pazzaglia FJ, Gardner TW (1993) Fluvial terraces of the lower Susquehanna River. Geomorphology 8:83–113

Peterjohn WT, Correll DL (1984) Nutrient dynamics in an agricultural watershed: observations on the role of a riparian forest. Ecology 65:1466–1475

Poff NL, Allan JD, Bain MB, Karr JR, Prestegaard KL, Richter BD, Sparks RE, Stromberg JC (1997) The natural flow regime: a paradigm for river conservation and restoration. BioScience 47:769–784

Riding JB, Kyffin-Hughes JE (2006) Further testing of a non-acid palynological preparation procedure. Palynology 30:69–87

Stanford SD, Ashley GM, Brenner GJ (2001) Late Cenozoic fluvial stratigraphy of the New Jersey Piedmont: a record of glacioeustasy, planation, and incision on a low-relief passive margin. J Geol 109:265–276

Stuiver M, Reimer PJ, Bard E, Beck JW, Burr GS, Hughen KA, Kromer B, McCormac G, van der Plicht J, Spurk M (1998) IntCal98 radiocarbon age calibration, 24,000–0 cal BP. Radiocarbon 40:1041–1083

Tomlinson J (2003) Land use and sediment yield: an assessment of Swarr Run, Lancaster, Pennsylvania. Senior Independent Thesis, Department of Geosciences (now Earth and Environment), Franklin and Marshall College, Lancaster, Pennsylvania

Trimble SW (1997) Contribution of stream channel erosion to sediment yield from an urbanizing watershed. Science 278:1442–1444

Voli M, Merritts D, Walter R, Ohlson E, Datin K, Rahnis M, Kratz L, Deng W, Hilgartner W, Hartranft J (2009) Preliminary reconstruction of a pre-European settlement valley bottom wetland, southeastern, Pennsylvania. Water Res Impact 11:11–13

Vörösmarty CJ, Sahagian D (2000) Anthropogenic disturbance of the terrestrial water cycle. BioScience 50:753–765

Walter RC, Merritts DL (2008) Natural streams and the legacy of water-powered mills. Science 319:299–304

Watts WA (1979) Late quaternary vegetation of central Appalachia and the New Jersey coastal plain. Ecol Monogr 49:427–469

Williams JW, Shuman BN, Webb T III, Bartlein PJ, Leduc PL (2004) Late-quaternary vegetation dynamics in North America: scaling from taxa to biomes. Ecol Monogr 74:309–334

Wolman MG, Schick AP (1967) Effects of construction on fluvial sediment, urban and suburban areas of Maryland. Water Resour Res 3:451–464

Zedler JB (2000) Progress in wetland restoration ecology. Trends Ecol Evol 15:402–407

Zedler JB (2003) Wetlands at your service: reducing impacts of agriculture at the watershed scale. Frontiers Ecol Environ 1:65–72

Chapter 4
A Paleoecological Perspective on Wetland Restoration

Christopher J. Williams

Abstract Paleoecological investigations of wetland sedimentary deposits offer the possibility of obtaining accurate reconstructions of base line conditions in the past. Plant remains, such as leaves, seeds, fruits, wood, and pollen, provide a window of variable temporal and spatial resolution into past environmental conditions at a particular site. These archives of physical and biological wetland ecosystem characteristics, if preserved, may be exploited to reconstruct the plant community at a single point in time. Moreover, changes in past plant community composition, hydrology, and the dynamics of wetland ecosystems through time may be better understood. This paper reviews the range of paleoecological information archived in wetland sedimentary deposits that may be understood in the restoration science context. This type of information gleaned by applying paleoecological techniques should provide reasonable targets for restoration ecologists working to improve the quality and quantity of ecosystem functions and services in wetlands.

Undisturbed wetlands, by virtue of their position on the landscape and the specific properties of the plants that populate these ecosystems, archive proxies of past environmental information within their soils and sediments. In temperate North America the geomorphic settings for wetlands typically vary from locale to locale, including depressional, riverine or floodplain, and lacustrine, among others. Yet all wetlands share a common attribute; the inundation of soil by water at a duration or frequency that is sufficient to exclude flora and fauna intolerant of such conditions. Although this ecosystem trait provides wetland scientists a means to mechanistically define a wetland, from a paleoecological standpoint there is a more important consequence of hydric soil conditions; plant and animal remains are often preserved in the wet, oxygen-deficient soils. Moreover, most wetlands are depositional environments; places where sediment accumulation exceeds erosion, thereby providing

C. J. Williams (✉)
Department of Earth and Environment, Franklin and Marshall College,
Lancaster, PA 17603, USA
e-mail: chris.williams@fandm.edu

B. A. LePage (ed.), *Wetlands,*
DOI 10.1007/978-94-007-0551-7_4, © Springer Science+Business Media B.V. 2011

the potential for the accumulation of long and nearly continuous depositional re-cords of such remains in these basins. Plant remains, such as leaves, seeds, fruits, wood, and pollen, provide a window of variable temporal and spatial resolution into past environmental conditions at a particular site. These archives of physical (abi-otic) and biological (biotic) wetland ecosystem characteristics, if preserved, may be exploited to reconstruct the plant community at a single point in time, but more importantly, changes in past plant community composition, hydrology, and the dy-namics of wetland ecosystems through time may be better understood. This infor-mation is valuable to those who are interested in studying the ecological dynamics of the past, such as paleoecologists, and to those who have more applied goals, such as practitioners of wetland creation, enhancement, and restoration.

The objectives of wetland restoration efforts may include a desire to restore spe-cific ecosystem function and services, the restoration of endangered species, and the re-establishment of wildlife habitat to name a few (Ehrenfeld 2000a). Goal-oriented restoration was driven early on by one overarching goal—the return of an ecosystem to a close approximation of its condition prior to disturbance (National Research Council 1992). However, as noted by Cairns and Heckman (1996) this im-mediately raises the question as to what frame of reference or benchmark does one use to establish the pre-disturbance condition? Moreover, others have demonstrated that restoring an ecosystem to a prescribed composition is difficult because the as-sumption of predictable community assembly is often not valid and restoration can result in unpredictable outcomes (Gwin et al. 1999; Magee et al. 1999; Zedler 2000; Magee and Kentula 2005). Such results led Hilderbrand et al. (2005) to go so far as to declare such prescriptive restoration approaches to be a so-called restoration myth (i.e., their "Carbon Copy" myth) in that creating or restoring an ecosystem that is a copy of an idealized state with a single endpoint is at odds with the dynamic nature of ecosystems. Some have called for a move away from the restoration of a system with a static community composition to a restoration that incorporates a more dynamic view of community composition (Walker et al. 2007).

Nevertheless, there is general consensus that for the restoration of some ecosys-tems, it is desirable to have adequate knowledge of the pre-disturbance conditions as a starting point for a restoration design. Nearly all restoration schemes call for some knowledge of ecosystem community composition when developing a resto-ration design. This might include knowledge about, and a focus on, species-level dynamics (Ehrenfeld 2000a), or higher order interactions, such as community-scale interactions and structural information (Hobbs and Norton 2004). And, for those who seek to re-establish pre-disturbance sucessional processes, knowledge of the pre-impact ecosystem composition, structure, and functioning is essential (Society for Ecological Restoration International Science and Policy Working Group 2004).

The lack of information regarding pre-disturbance ecosystem conditions is viewed as a problem common to many restoration projects (Clewell and Rieger 1997). The common sources of baseline or reference information also vary in tem-poral and spatial resolution. In some cases, baseline conditions may be gleaned from historical ecological inventories, photos, and herbarium and museum speci-mens (Kowalski and Wilcox 1999; Pellerin and Lavoie 2003; Bowen et al. 2004;

Leck and Leck 2005), or in some cases, even remnants of the site to be restored (Lavoie et al. 2001). Even in areas with adequate historical records though, these may already bear the fingerprints of human activity (Black and Abrams 2001), and so it is desirable to utilize multiple methods to overcome the limitations of a single source of reference data.

Wetland restorationists and managers are also increasingly recognizing that ecosystem characteristics can be highly variable in time and space (Landres et al. 1999; Hobbs and Harris 2001). Some ecosystems are now managed for variability (Keane et al. 2009) and it is in this sense that paleoecology can contribute very meaningfully to restoration science. Because paleoecological techniques provide reference information at a range of both spatial and temporal scales, they may provide a means to represent the dynamic character of ecosystems as they vary over time and across landscapes (Swanson et al. 1994). This is compatible with the desire to understand the historical range of variation in ecosystem properties (White and Walker 1997) and incorporate this information into the design of restoration projects (Palmer et al. 1997). In the context of wetland restoration, the range of Holocene (approximately the past 10,000 years) variation probably provides the most useful temporal context within which to evaluate the variation in ecosystem traits given that there appear to be relatively long periods (thousands of years) of environmental stability in these records (Wanner et al. 2008). At this time scale, paleoecological evidence such as fossil plant remains may provide several pieces of information useful to the development of a restoration design that we can manage and restore in ways that allow for some amount of variation in ecosystem properties. This is particularly important when planning restorations to accommodate possible future climate change.

To this end, I review these sources of information in the restoration science context. The goal here is not to present detailed methodologies for which several texts exist (Birks and Birks 1980; Berglund 1986; Jones and Rowe 1999). Rather, this chapter provides restoration practitioners with an overview of the type of information that may be gleaned from paleoecological investigations of sedimentary deposits, if such methods are applied. As a practioner of paleoecology there are numerous applied aspects of the discipline that could be useful in a restoration sense. This long-term perspective improves our understanding of the baseline conditions; the state of an ecosystem prior to major anthropogenic disturbance, which is a common goal of ecosystem restoration. Obtaining accurate reconstructions of baseline conditions in the past should provide reasonable targets for restoration ecologists working to improve the quality and quantity of ecosystem functions and services in wetlands. I begin with a review of the spatial and temporal aspects pertinent to the use of such data.

Temporal and Spatial Aspects of Paleoecological Data

The remains of plants and animals preserved in wetland or floodplain environments may span a wide range of spatial and temporal scales. Thus, the application of paleoecological data to restoration projects then depends on an understanding of the amount of ecological time and space represented by a fossil deposit since these quantities may vary depending on the nature of the depositional environment and the type of material preserved (Behrensmeyer and Hook 1992). Bennington et al. (2009) recently reviewed the issue of relating paleoecological data to modern (neo-ecological) studies and they emphasize the important point that a disconnect between paleoecology and neoecology does not arise due to differences in the nature or completeness of the data, but rather, that these two data sets commonly represent different temporal scales. In this context, time averaging is an important concept to consider in understanding the temporal scale of paleoecological data.

The interval of time represented by an assemblage of plant material may be variable because the plant remains of several generations of plants may be deposited within a single stratigraphic horizon. This is due to the fact that sediment accumulation rates are generally low relative to generation or population turnover times and associated deposition of plant materials. This process, referred to as taphonomic time-averaging (Behrensmeyer and Hook 1992), will determine how finely ecological time can be resolved in a particular fossil deposit. DiMichele and Gastaldo (2008) liken the whole process to compiling a photographic record using a camera with a variable speed shutter. Some fossil deposits may represent a brief moment in time, similar to a photographic snapshot of limited ecological time and space (Fig. 4.1). For instance, envision a floodplain forest floor, covered with fallen leaves and seeds produced by the vegetation growing at the site. If this floodplain surface were to be rapidly buried and preserved in sediments (Shure et al. 1986; Clague et al. 2003), any plant material incorporated into these flood sediments would then represent a snapshot of the vegetation at the time of the flood. Thus, the plant fossil record preserved would be akin to a photograph taken at a relatively high shutter speed. In contrast, if one were to envision the preservation of plant litter in a small depressional peatland, such as a kettle hole bog (Fig. 4.1) that was slowly filling with organic matter, then the entire deposit may represent 1,000–10,000 years; analogous to taking a long-exposure photograph. Over this long exposure time some plant species may migrate into the wetland and then go locally extinct as wetland hydrology varied with long-term climate change. These fleeting wetland residents would leave only a minor trace, if at all, of their existence in the larger sedimentary record that may be too short to resolve in the fossil record.

The temporal resolution of a Holocene floodplain or wetland deposit may vary widely. A single layer of fossil plant material in a floodplain deposit may represent 1–100 years of accumulation (Fig. 4.1). Channel fill deposits that form on a floodplain may time-average fossil assemblages at 100–10,000 years, whereas channel lag and bar deposits may time-average fossil material at even longer intervals (Behrensmeyer and Hook 1992). However, sedimentary deposits where accumulation rates are relatively high will offer the best resolving power (Pederson et al. 2005) with some wetlands providing a high temporal sampling resolution, such as less than 10 years per 1 cm of thickness (Yu 2006). Thus, the potential for very high-resolution reconstruction of vegetation composition may exist, although cases of such exceptional temporal resolution are probably the exception rather than the rule.

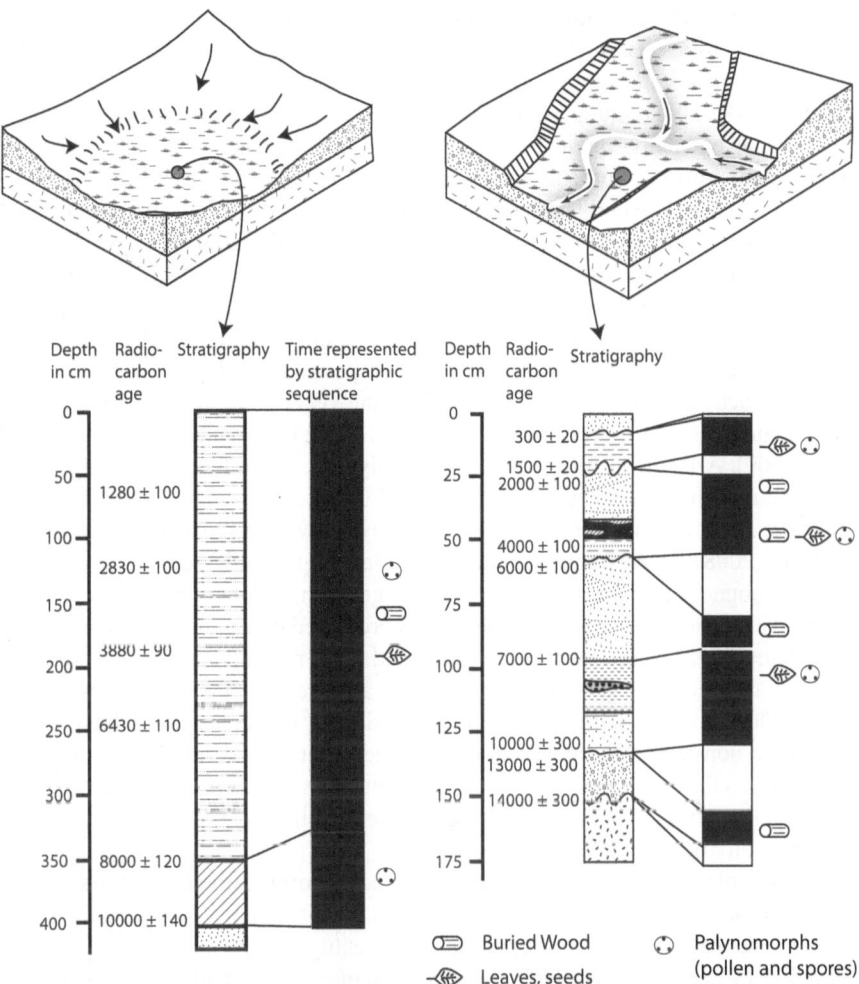

Fig. 4.1 Hypothetical wetland deposits and stratigraphic representations of deposits and time recorded by such deposits. For a small depressional wetland (*top left*), peat deposition and accumulation may occur continuously and thus the time represented by the deposits may be close to complete (*black regions* of second column) even as the system transitions from a fen to a bog. Pollen and macrofossils may be found throughout the peat deposit. In contrast, a floodplain deposit (*top right*) may transition from periods of deposition (*black regions* of second column) to periods of erosion (*white regions* of second column). Thick sedimentary deposits may represent short periods of time whereas some thin deposits may represent long periods of time. In both systems, periods of deposition that preserve fossils may be taphonomically time averaged

Whereas the temporal resolution of fossil deposits may be quite variable and not always well known, the spatial resolution of a fossil assemblage may be more confidently understood. As noted by Bennington et al. (2009), a critical question to address is: How much ecological space is captured in an individual sample?

The answer to this question relies largely on a number of actual studies of living plant communities (Burnham et al. 1992; Wing et al. 1992 and references therein). By and large, repeated study indicates that the great majority of plant species with readily preservable remains are in fact represented in their local fossil assemblages. In terms of paleoenvironmental information useful for understanding restoration reference conditions, autochthonous (no transport or *in situ*) to parautochthonous (limited transport) assemblages are the most desirable types of assemblages to study because they represent fossils sourced from constrained spatial areas. Such assemblages most often are found within floodplain lakes (Rich 1989; Gastaldo et al. 1989, 1998), although fluvial (river or stream) channel lag and bar deposits also may contain parautochthonous assemblages depending on the fluvial dynamics and aerial extent of the vegetation across the landscape. In general, plant parts preserved in fluvio-lacustrine settings are derived from the vegetation growing directly adjacent to and fringing the water body (Wing et al. 1992; Baker and Drake 1994; Baker et al. 2000). After deposition, there is generally little or no transport of plant parts into or out of the area of burial, although pollen and spores may be transported in, from outside of the immediate area because they are wind dispersed (DiMichele and Gastaldo 2008). Buried autochthonous litter layers generally record the composition and abundance patterns of the source vegetation, although rare and evergreen plants within the community may be underrepresented or absent in the fossil record (Burnham et al. 1992). Herbaceous plants and ferns may also be underrepresented in the plant litter record because they do not abscise their leaves (Ferguson et al. 1999). On the other hand, studies of deposition in modern settings have shown that some non-locally (allochthonous) derived plant fossils may be incorporated into deposits that are dominated by parautochthonous fossils (Gastaldo et al. 1987). Nevertheless, these in many cases these may be identified as transported material as they typically bear evidence of abrasion.

The samples of the plant fossil record that may inform restoration efforts are then subject to time averaging, and may be sourced from areas of varying spatial extent and in some cases may under represent the diversity of rare species. Nevertheless, these biases occur in a relatively consistent manner in wetland environments, and may therefore be accounted for to maximize the utility of information recovered from fossil deposits.

Plant Macrofossils

In general, plant macrofossils are preserved organic remains large enough to be visible without a microscope or hand lens (Fig. 4.2a, b). In most cases, macrofossils are defined operationally on a size-class cutoff for materials retained on a particular sized-sieve (greater than 200 μm mesh opening) with the intent of capturing materials such as seeds and wood fragments while washing away the finer materials that may obscure the samples of interest. I will review at each of these sources individually.

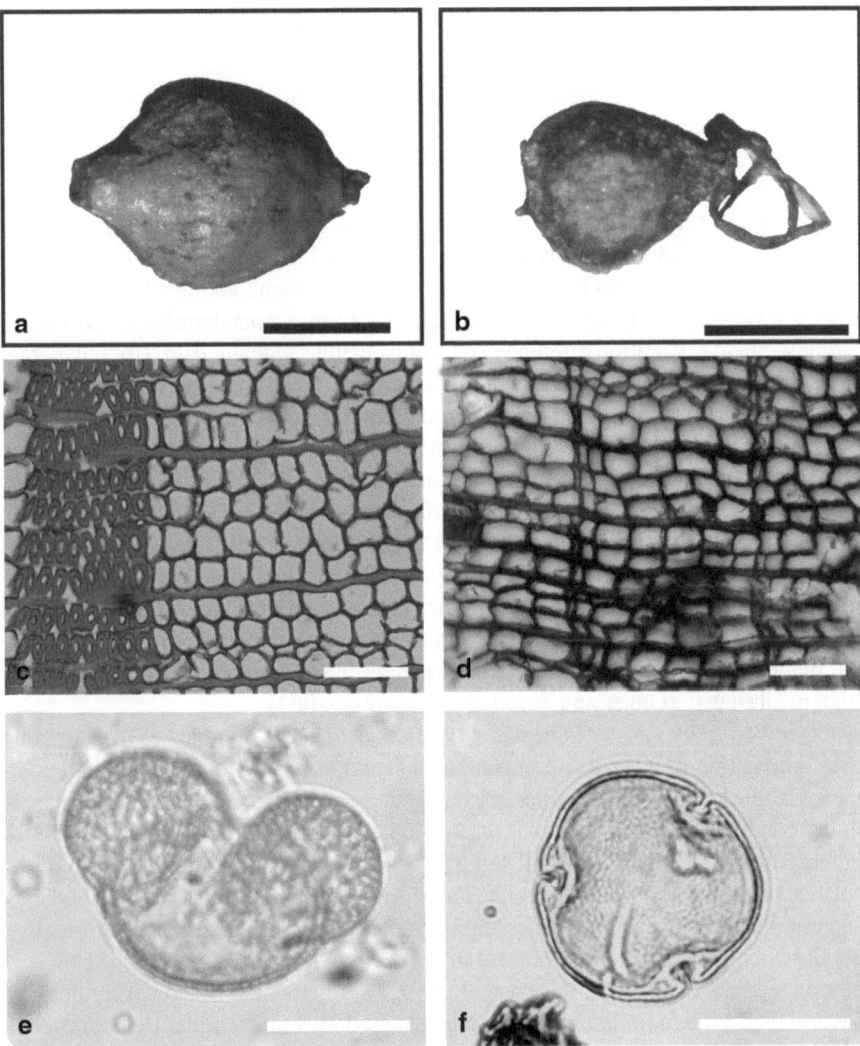

Fig. 4.2 Representative macro- and microfossil materials such as seeds, wood, and pollen recovered from sedimentary deposits. **a** Subfossils of *Carex* (sedge) and **b** *Eleocharis* (spikerush) seeds recovered from late Holocene wetland deposits at Dillerville wetland near Lancaster, Pennsylvania. Scale bars in **a** and **b** equal 1 mm. **c** Transverse thin section of subfossil Taxodiaceae wood (*cf. Metasequoia* [dawn redwood]) recovered from a late Holocene (*ca.* 1,400 ^{14}C years BP) peat deposit in near Lichuan, China. **d** Transverse thin section of subfossil Pinaceae wood (*cf. Larix* [larch]) from a late Holocene peat deposit at Labrador Hollow, New York. Scale bars in **c** and **d** equal 100 μm. **e** Equatorial view of a bisacate *Pinus* (pine) pollen grain. **f** Polar view of a tricolporate *Tilia* (linden) pollen grain. Scale bars in **e** and **f** equal 25 μm

Seeds

The study of wetland seedbanks is fairly well established and has been studied in depressional wetlands, peatlands, tidal freshwater wetlands, floodplains, and lacustrine settings (van der Valk et al. 1992; Peterson and Baldwin 2004; Egawa et al. 2009; Neill et al. 2009). The soil seedbank is an archive of the recent and past vegetation of a wetland area. In a restoration context, it is an indication of the species composition at a site when recent soils are evaluated. However, there may not be a strong similarity between the composition of the seedbank and the composition of standing vegetation; a caveat that should be taken into consideration when formulating wetland restoration reference conditions using seedbank data. Hopfensperger (2007) conducted a meta-analysis of the data in the literature to determine the similarity between seedbank composition and standing vegetation in wetlands. Using Sørenson's index of similarity as a metric of floristic similarity between seedbank and standing vegetation, she found that across 42 studies of wetlands the average similarity index was 47% and ranged from about 17–79% (Hopfensperger 2007). The seeds of early succession or pioneer species may persist in soils even though environmental conditions such as light availability or thick litter preclude their germination and thus result in relatively low similarity values. High similarity values may be explained by the short dispersal distance or the clustering of seeds near parent plants. Likewise, wetlands with annual-dominated plant communities are often high in similarity (Ungar and Woodell 1996; Jutila 2003).

As reviewed by van der Valk et al. (1992), there are two approaches to estimating the composition of a seedbank: mechanical separation of seeds and seedling emergence studies. Emergence studies are based on soil collected at a site that is placed under conditions suitable for seeds to germinate. The number and type of plants that emerge from these incubated soil samples then provide a minimum estimate of the species composition of the seedbank. This approach is more frequently applied than separation studies because it is less labor intensive and provides some information about which plant species are present in the seedbank for *in situ* restoration of vegetation (Bakker et al. 1996; Vécrin et al. 2002; Middleton 2003). However, there are several well-known limitations to emergence studies. For example, Benthardt et al. (2008) found that separation methods provided a better quantitative description and seed recovery of four wetland-plant species (*Carex bohemica* [Bohemian sedge], *Coleanthus subtilis* [moss grass], *Elatine hexandra* [six-stamened waterwort], and *Eleocharis ovata* [ovate spikerush]) recovered from pond sediments in Austria. The averaged detection failure across the four study species was almost 90% with the seedling-emergence method. However, Poiani and Johnson (1988) found that the emergence method gave an accurate assessment of the number of species and the relative abundance of seeds present compared with actual identification of seeds. On the other hand, Price et al. (2010) had lower recovery of seeds from an ephemeral wetland pool using separation techniques as compared to emergence studies of more deeply buried seeds. Nevertheless, seed dormancy and a failure to meet the specific requirements of all of the species present in the seedbank are likely to cause an underestimate of the species composition of the wetland soil seedbank (Roberts 1981; Gross 1990; van der Valk et al. 1992). An underestimate of seedbank species composition may also result from differences in the duration of seed viability among the seeds of the different wetland plant taxa (Price et al. 2010). For example, van der Valk et al. (1999) demonstrated that the seeds of several species of *Carex* lose their viability (ability to geminate) in less than 1 year and may be poorly represented in seedling emergence studies compared to seeds that possess greater viability over longer time spans. Given that variable results have been generated using different methods it may be appropriate to utilize both seedling emergence and mechanical separation studies depending on the scope of the study (ter Heerdt et al. 1996).

Using mechanical separation of seeds followed by hand sorting and identification under a microscope is less frequently applied, but offers the added advantage of not being influenced by differences in germination requirements of the studied species. Moreover, the viability of seeds may on the order of years to decades (Leck and Schütz 2005). Thus, physically separating and identifying the seeds allows one to reconstruct a more complete species composition list that includes non-viable seeds, which is an important consideration when trying to reconstruct past patterns in vegetation spanning millennia. Aside from the time consuming nature of the enterprise, one concern associated with using mechanical separation as a means of categorizing buried seeds is that small seeds (less than 2 mm) may be overlooked or lost during sieving due to use of too large a sieve size (Beaudoin 2007; Price et al. 2010). It is worth noting that whereas seedling emergence studies may underestimate the composition of the wetland soil seedbank, mechanical separation methods may overestimate the population size of viable seeds. Recognition of this bias is important for those interested in using *in situ* seedbanks, as opposed to sowing seed, as source material for restoration efforts. However, this should not be a major concern when the overall goal of the seedbank study is to develop a simple account of plants that once grew at a site.

Identification of seeds requires comparison to modern reference plant material keyed in the field or obtained from herbaria sheets, as the available photographic atlases are limited (Martin and Barkley 1961; Montgomery 1976; Berggren 1969; Schopmeyer 1974; Flood 1986; Hurd et al. 1994, 1998). Following identification, a so-called "seed flora" may be constructed for particular sedimentary intervals allowing for simple presence-absence type data to be recorded for some deposits. This was the approach of Voli et al. (2009) in their analysis of approximately 3,200-year-old seeds from a buried Holocene wetland. Seeds that are separated from a known volume of sediment or peat (e.g., 10–100 cm^3) can then be quantitatively analyzed or compared between the stratigraphic intervals that were sampled. The data can then be presented graphically (Fig. 4.3; Yu et al. 1996; Magyari et al. 2001; van der Putten et al. 2009), using Tilia (Grimm 1993) or C2 (Juggins 2007) software. When paired with information on other macrofossil and/or sedimentology data, the seed data may provide information regarding the response of a wetland system to environmental change. This technique was used by Payette and Delwaide (2004)

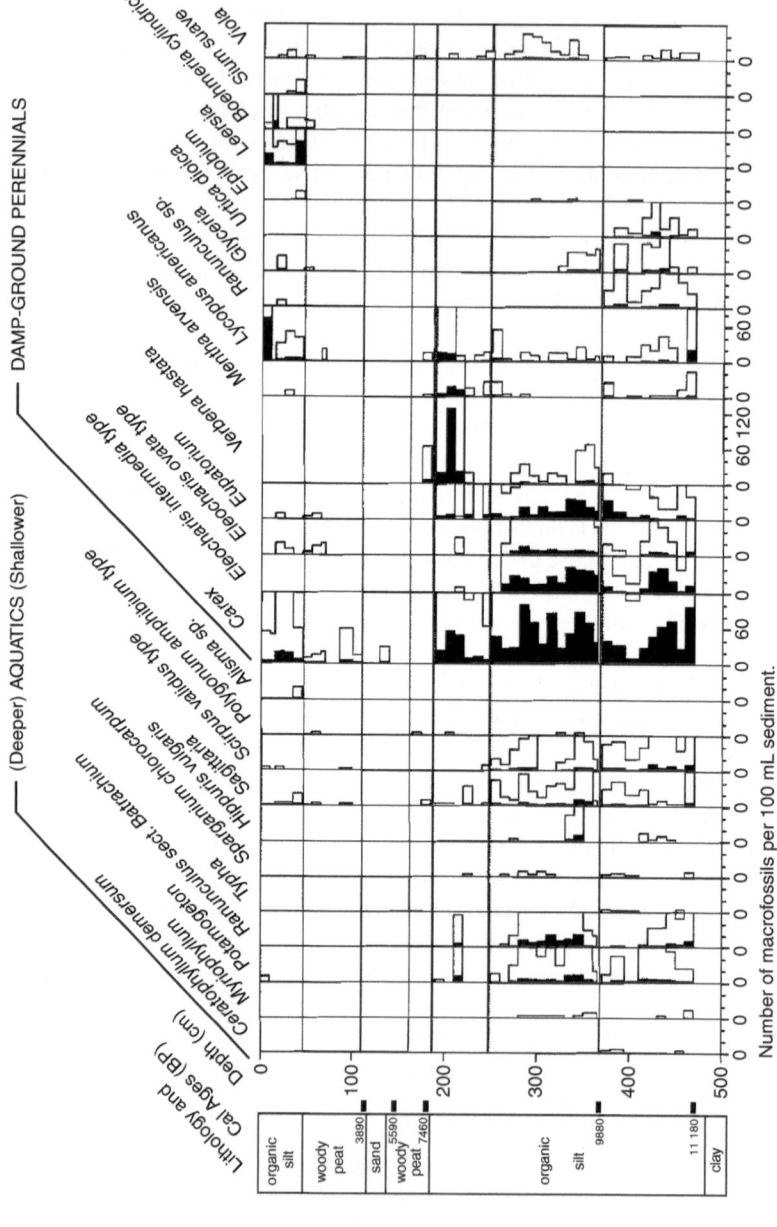

Fig. 4.3 Graphical representation of seeds, fruits and achenes as presented by Yu et al. (1996). (Reprinted by permission, © 2008 NRC Canada or its licensors)

to study environmental changes in subarctic wetlands associated with moisture and temperature changes over the past 1,500 years.

Wood

There is growing interest in the importance of woody vegetation in riparian areas and the geomorphologic impact of woody debris sourced from such vegetation (Gurnell et al. 2005). For example, standing trees on floodplains increase channel roughness during overbank flows, thereby resulting in a decrease in the erosive actions of floods and an increase in the retention of material being transported by flood waters (Anderson et al. 2006). Also, woody debris deposited in channels provides many important functions in stream ecosystems such as sediment retention and habitat formation (Abbe and Montgomery 1996; Bilby and Bisson 1998; Gurnell et al. 2005). However, as noted by Rheinhardt et al. (2009), reference information on riparian zone forests is sparse for many biogeographical regions, especially the composition and structural data that would provide useful targets for restoration.

This is particularly true in regions where settlement associated with agriculture began centuries ago and the remnant vegetation in what appear to be relatively undisturbed ecosystems may actually be impacted by landuse practices in adjoining areas (Bedford 1999; Houlahan and Findlay 2004). For example, nutrients may migrate into riparian areas adjacent to agricultural ecosystems and facilitate the growth of plants that might not otherwise be able to compete under nutrient poor soil conditions. Many riparian and floodplain areas in the eastern United States also became filled with sediment as a result of high sediment erosion rates associated with land clearing and stream impoundment. This resulted in the pre-settlement floodplain surface becoming entombed in the deposited sediments (Ruhlman and Nutter 1999; Walter and Merritts 2008; de Wet et al. 2011). Wood buried beneath these sediments may then be an important source of information for guiding restoration designs and establishing pre-disturbance reference conditions.

Disarticulated remains of woody plants now exposed by stream channel incision or restoration associated excavation activities can be used to reconstruct the structure (size and distribution) of riparian vegetation. Under the appropriate conditions, *in situ* tree stumps preserved in growth position (Fig. 4.4a, c) can be used to gain a more refined view of the structure and developmental history of forested wetlands, thereby providing much needed pre-disturbance riparian compositional and structural information. Despite the extensive study of fossil wood in paleoecological studies that span large amounts of geologic time, there has been only limited application of this proxy of past vegetation to guide restoration designs (Walker et al. 2003). In fact, we may know more about the local-scale composition and structure of forests that are many thousands or millions of years old from studies of buried wood (Pregitzer et al. 2000; Williams et al. 2003, 2008; Yamakawa et al. 2008) than we do about forests that existed at the time of European settlement in the eastern United States which were cut and cleared. Therefore, if adequate exposures are

Fig. 4.4 Pre-European settlement forest remains exposed
in the stream channel of
Pennypack Creek, northeast
Philadelphia, Pennsylvania. **a**
Oblique view of stream channel with numerous stumps
(select stumps indicated by
arrows) exposed by stream
incision following a mill
dam breach less than 100 m
downstream from the area in
the photograph. **b** Laminated
sediments deposited behind
the mill dam impoundment
with plant macrofossil material indicated by *white mark-
ers*. **c** Large (approximately
70 cm diameter hollow stump
and log exposed in the stream
channel)

available and the tools of plant paleoecology are brought to bear on more recent
buried wood deposits it would be possible to gain a higher resolution view of forest
composition.

    To first order, the most basic information that may be extracted from buried wood
is its taxonomic identity. Wood identification is based primarily on the cellular or

anatomical structure of the wood (Fig. 4.2c, d). Observation of microscopic wood anatomy is often sufficient to identify the wood to the genus level and sometimes to the species level with careful comparison of the unknown wood to samples of botanically authenticated samples (Kellogg and Rowe 1981; Tennessen et al. 2002). The physical properties such as color, odor, and density that are useful for modern wood identification may also be useful in exceptionally preserved fossil wood. Wood identification may be facilitated by utilization of numerous wood anatomy atlases and dichotomous keys (Hough 1957; Panshin and deZeeuw 1980; Schweingruber 1990; Hoadley 1990). More recent online resources such as the "*Insidewood Database*" (http://insidewood.lib.ncsu.edu/search), the "*Wood Anatomy of Central European Species*" database (www.woodanatomy.ch), and the "*Xylem Databank*" (http://www.wsl.ch/dendro/xylemdb/index.php) offer images of tree rings and wood anatomy that prove to be a valuable resource in making wood identifications. The Center for Wood Anatomy Research at the United States Department of Agriculture, Forest Products Laboratory in Madison, Wisconsin, USA also offers limited wood identification services and houses an extensive research wood collection to aid in wood identification.

As with other macrofossil material types, the depositional context in which the fossil wood is emplaced is important for determining the provenance of the wood and drawing ecological inferences. Even in dynamic fluvial environments wood is not particularly mobile; large pieces may have general residence times of three to five decades in active channels (Hyatt and Naiman 2001) and even small pieces (less than channel width) of woody debris may not travel far (Jones and Smock 1991). For example, Gurnell et al. (2002) studied wood movement in the 9 m wide channel of Mack Creek in a conifer dominated catchment in Oregon, USA. Less than 1% of the logs in Mack Creek moved in most years and even during a flood (25 year recurrence interval), 89% of the wood pieces remained in their original positions, and only 11% of the pieces moved more than 10 m. However, samples emplaced in bar and lag deposits are more likely to represent vegetation from a greater spatial area along the reach of the stream. Wood on the floodplain may travel along the floodplain during inundation, thus forming parautochthonous deposits, but generally only low percentages of wood deposited on the floodplain get transported into the channel (Jones and Smock 1991). Autochthonous wood assemblages are most likely to represent the canopy and even sub-canopy composition of the woody vegetation that occupied the wetland or riparian floodplain environment.

Autochthonous assemblages of woody debris (Fig. 4.4a, c), where stems and logs are preserved *in situ*, may allow for relatively complete reconstruction of both the taxonomic composition and the structural attributes of a forest such as tree spacing and stem size. Stumps preserve features that allow the determination of different genera from wood anatomy, the original basal diameters of the trees, and the spatial distribution of individuals. Logs retain dimensional information such as stem size, taper, and branching that can provide information about vertical forest structure. Longitudinal sections of tree trunks contain buried knots that preserve the history of branching and can be used to interpret stand dynamics (Larson 1963; LeBlanc 1990). Treetops provide information on the vertical distribution and size of live branches. Numerical techniques, borrowed from neoecological studies can be applied to paleoecological studies in some situations (Williams 2007). When integrated, all of these sources of data may be informative about the structure of pre-historic ecosystems.

For example, Kooistra et al. (2006) examined a 1,270 m² area of buried *in situ* forest from a late Holocene peat deposit (approximately 1,850–1,650 calibrated years before present [cal. BP]) in the Netherlands. In their study more than 600 fragments of wood were mapped and more than 500 of these samples were taxonomically identified. Measured logs were used to reconstruct tree height and dendrochronology was used to reconstruct the age structure of the buried forest. They described recovered trunks that were generally straight without massive limbs close to the stem base which led them to conclude that the trees (mainly oak [*Quercus*] and ash [*Fraxinus*]) had grown in a closed stand. A similar approach was taken by Pregitzer et al. (2000), albeit for a much older forest (approximately 9,900 cal. BP) in northern Michigan. They were able to use stems preserved in growth position to reconstruct the size structure of the forest as well as the general spacing of the trees. Although the great age of the buried forest makes it less useful for providing reference conditions for the restoration of modern forests, it still demonstrates the level of detail that can be reconstructed when an autochthonous assemblage is analyzed. In some cases, impoundment of the stream behind milldams may result in the flooding, burial, and preservation of the remains of floodplain forests associated with early land clearance in eastern North America. Williams (unpublished data) mapped the remains of an *in situ* colonial age floodplain forest in the Pennypack Creek area of Philadelphia, Pennsylvania, USA (Fig. 4.4). Spatial analysis of the stumps of trees exposed by stream incision following a dam breach reveals the size, structure, and spacing of the colonial age trees of this forest and gives some insight into the nature of the forests associated with the early colonial landscape in this location.

Palynology

In contrast to the aforementioned sources of paleoecological information, the study of palynomorphs (pollen and spores) has been more broadly applied to provide context for restoration and conservation projects, although at slightly lower spatial resolution than the macrofossil assemblages provide (Brown and Pasternack 2005; Burney and Burney 2007; van Leeuwen et al. 2008; Jackson and Hobbs 2009; Brown 2010). The broad appeal of using fossil pollen and spores to gain insight into pre-disturbance conditions corresponds to their high degree of decay resistance, which leads to good preservation potential in late Quaternary sedimentary and wetland deposits (Fig. 4.2e, f). Pollen may be found in abundance in the accumulated sediments of lakes, wetlands, and estuaries, as well as in soils (both contemporary and paleosols), and alluvial (entrained material deposited by flowing water) and colluvial (slope material deposited by gravity flow) deposits. Another reason for the broad application of pollen as a proxy for past vegetation is that the source plants can usually be identified to the genus level and sometimes the species level if enough morphological variation between pollen grains exists. Thus, changes recorded in the pollen record can be used to infer changes in landuse, climate, hydrology, and disturbance regimes on time scales of 100–1,000 years. Perhaps the greatest limitation to the application of pollen to answer questions in restoration science is the fact that pollen analysis is labor intensive and requires specialized training in order to confidently assign fossil pollen to specific taxa.

The temporal and spatial resolutions at which changes in the predominant vegetation can be detected in the pollen record again depend on the depositional environment and rate of sedimentation at a site. A high temporal resolution may be important in certain restoration applications, for example trying to discern the response of local vegetation to historic short-term changes in the environment such as drought, flooding, or fire or landuse. Because the temporal resolution of the pollen record is generally affected by the vertical thickness of the sediment sample to be studied, the rate of sediment accumulation in the basin, and the age-depth model employed, some depositional environments may be more suitable than others for high-resolution work. In this case, a continuous time series without gaps in the sampling and with no bioturbation will improve the temporal resolution. In some cases, annually laminated lake sediments may provide the best temporal resolution if each lamina is thick enough to provide enough material for pollen analysis (Peglar 1993). In other regions, peat deposits may accumulate more rapidly than lake sediments thereby providing better temporal resolution for pollen studies (Bennett and Hicks 2005; Barnekow et al. 2007). However, numerous published papers present vegetation reconstructions based on pollen data at coarser temporal resolutions, mainly at the centennial scale, and these may be suitable for work in restoration science if the goal is to understand the local or regional history of vegetation at a particular site.

Pollen recovered from lake sediments is usually taken to represent a mosaic of vegetation growing in a source area up to tens of kilometers in distance from a site. Pollen loading at this scale is considered to be the so-called background pollen loading or regional pollen component. That is to say, within a limited geographic area pollen loading does not vary between similarly sized sampling sites on the landscape (Sugita 1994). Long-term vegetation changes are best resolved by pollen-based reconstructions that occur at the local to sub-regional scale (Jacobson and Bradshaw 1981). For example at the smallest scale, Calcotte (1995) and others have demonstrated that the fossil pollen recovered from small isolated wet areas within a forest or small basin can give a stand-scale view of vegetation (Sugita 1994). In this setting, local pollen contributes about 45–50% of the pollen input to small forested hollows in mixed hardwood conifer forests, which is enough to characterize the contributions of the major plant taxa that grew within about 50–100 m of the sample point (Calcotte 1995, 1998; Sugita 1994). Thus, pollen studies conducted in basins of differing sizes may afford an opportunity to study changes in both local and regional vegetation and the response of local vegetation to different environmental forcing factors (Schauffler and Jacobson 2002). More often than not though, pollen records collected at coarser spatial and temporal resolutions are more readily available for a particular region and these data may serve the needs of the restoration project. For example, the *Global Pollen Database* provides pollen data via the Internet (http://www.ncdc.noaa.gov/paleo/ftp-pollen.html) and allows data from individual sites to be accessed for analysis. In most cases, such data can be used to assemble a basic list of the dominant pre-disturbance plant taxa in a particular area.

The application of palynology to restoration sciences in general has been applied in varied ways in several instances in published literature. van Leeuwen et al. (2008) used fossil pollen recovered from four wetlands in Galápagos, Ecuador to determine that six plant species that were thought to be non-native plants, so-called "doubtful natives" in the parlance of Jackson (1997), were in fact native to the Galápagos Archipelago. Their analysis of pollen and plant macrofossils suggests that all six pollen taxa were present thousands of years before the onset of human impact. Establishing this evidence for the native status of these plants is relevant to the issue of non-native species management and ecosystem restoration in the Galápagos Archipelago (Mauchamp 1997).

Brown (2010) utilized pollen data collected from sediments in a small forest hollow in southeast Wales, U.K. to establish a 2,000 year history of a 400-ha woodland and to use this knowledge to guide conservation and woodland restoration strategies in the region. The pollen data indicate that significant structural and compositional changes have occurred in the forested area during the last two millennia, which culminated in the planting of non-native conifers in parts of the woodland in the late nineteenth century. As a consequence of human impact, the present day woodland is dominated by planted non-native trees that bear little resemblance to the woodlands expected to occur naturally in the Welsh uplands (Brown 2010). Interestingly, Brown (2010) concludes that restoration to a pre-anthropogenic-disturbance state is probably unrealistic given that the woodland species represented in the fossil pollen record indicate many millennia of human management. Rather they suggest that restoration efforts focus on restoring the woodlands to only those habitats, shown through paleoenvironmental or historical data, to have occurred on the site prior to the introduction of non-native conifers. Thus, the restoration would result in woodlands that are similar in composition to that evident in the pollen record that came about due to traditional/historic forest management practices. An analogous situation may exist in some areas of eastern North America where pre-European settlement forest composition may bear the legacy of long-term active forest management by Native Americans. Abrams and Nowacki (2008) concluded that Native Americans managed the vast majority of vegetation in the eastern United States directly or indirectly using fire as a management tool. Moreover they posited that Native Americans directly promoted mast (nuts and acorns) and fruit trees by cultivation and planting them in the eastern forests (Abrams and Nowacki 2008 and references therein). This exemplifies the complex nature of human-vegetation interactions that may appear in the paleoecological record and highlights the problem of determining what a "natural" reference condition for ecosystem restoration might be. Moreover, paleoecological data, such as described in the aforementioned studies, indicate that the present day environmental conditions and even potential post-restoration conditions may preclude the re-establishment of a pre-disturbance flora and may suggest that a different or unique flora may result from restoration efforts.

Ehrenfeld (2000b) also provides an excellent example of the utility of paleoecological data in the context of wetland restoration in urban ecosystems. The Hackensack Meadowlands in New Jersey, USA is currently a brackish and saltwater marsh complex dominated by *Pharagmites australis* (common reed). The wetlands are surrounded by extensive urban development and are the focus of recent wetland restoration efforts (Tiner et al. 2002). The hydrology of the Meadowlands has been extensively altered with the largest impact resulting from decreased freshwater inflow due to impoundment of the Passaic and Hackensack Rivers and the intrusion of brackish water because of river dredging and rising sea level (Marshall 2004 and references therein). The net result has been an increase in the salinity of the Meadowlands that has resulted in the conversion of freshwater marshes into brackish and saltwater wetlands (Ehrenfeld 2000b). However, historical evidence indicates that freshwater-forested wetlands, dominated by *Chamaecyparis thyoides* (Atlantic white cedar), were an important part of the Meadowlands landscape (Sipple 1972). Analysis of pollen and buried wood collected from different locations within the Meadowlands (Heusser 1949, 1963; Carmichael 1980) confirm this notion and also indicate that open freshwater marshes, populated by a variety of graminoid (grasses) and cyperaceous (sedges) plants existed prior to anthropogenic disturbance, whereas *Phragmites* marshes dominate today. The local extinction of *Chamaecyparis* presents a major challenge to restoration ecologists given that it persists within a narrow set of environmental conditions (Laderman 2003; Mylecraine et al. 2005), which are not likely to be restored in the Meadowlands. This will preclude re-establishment of *Chamaecyparis* wetlands in this area. Indeed, this led Ehrenfeld (2000b) to conclude the wetlands such as the Meadowlands that are surrounded by extensive urban development may need to have a unique set of restoration benchmarks that are different from similar systems in less-developed areas.

Emerging DNA Techniques

The techniques described above draw on the traditional strengths of paleoecology and offer the potential for reconstructing the composition of wetlands prior to degradation and may very well provide some knowledge of the larger scale environmental conditions that defined past wetland ecosystems. Where do we go from here? One of the largest impediments to applied paleoecology is a lack of reliably identifiable remains in the sediments of interest. Even when remains such as wood fragments or pollen are found it may be difficult to identify these beyond the genus level. However, recent advances in molecular biology, particularly DNA analysis may provide a tool to overcome this barrier (see reviews of Gugerli et al. 2005; Schlumbaum et al. 2008). The concept of molecular identification, so-called "DNA barcoding," offers the prospect of rapid identification of fragments of organic matter of biological origin and has recently been applied to plants and fungi (Chase and Fay 2009). To be useful in a restoration context, fragments of DNA from plant and animal remains need to be recovered from wetland sediments or water (Ficetola et al. 2008). The DNA is then isolated and a standard region(s) of DNA is then sequenced and then matched with sequences of identified species available in publish databases (GenBank [http://www.ncbi.nlm.nih.gov/Genbank/]; Barcode of Life Data Systems [http://www.boldsystems.org/views/idrequest.php]). The best opportunity for successfully identifying plant remains to the species level will be in situations that involve geographically restricted sample sets such as studies focusing on plants from a given local area or ecosystem (Hollingsworth et al. 2009), which bodes well for the analysis of DNA in isolated depositional environments such as wetlands. Interestingly, owing to its speciose nature, the common wetland plant *Carex* has been studied in detail from a DNA barcode perspective (Starr et al. 2009).

Although still in its infancy, the application of DNA analysis to understand ancient assemblages of plants and animals has been undertaken by several researchers. Samples of DNA have been recovered from ancient permafrost and cave soils (Willerslev et al. 2003; Hofreiter et al. 2003), and wetland soils (Suyama et al. 2008) that offer the prospect for identifying and listing taxa that may have been present at a particular site. In one remarkable study, Willerslev et al. (2007) analyzed ancient DNA recovered from a deep ice core from south-central Greenland that may be between 450,000 and 800,000 years old. The DNA was assignable to trees typical of northern boreal forest ecosystems (alder [*Alnus*], spruce [*Picea*], pine [*Pinus*]) as well as a variety of herbaceous plants and provides evidence for reconstructing the composition of an ancient boreal forest in an area that currently resides under nearly 2 km of ice. Suyama et al. (2008) worked with more recent material to understand the Holocene population dynamics of a *Sphagnum fuscum* (rusty peat moss) in a Norwegian peatland. The analysis of ancient DNA has been extended to fossil plant pollen (Parducci et al. 2005; Bennett and Parducci 2006) and has also been recovered from both modern and ancient wood (Liepelt et al. 2006; Finkeldey et al. 2010). Although many limitations currently exist in the analysis of ancient DNA, the potential source material of ancient DNA from seeds, leaves, pollen, and wood is frequently encountered in wetlands and may hold great promise as a tool for reconstructing past wetland community composition, as these methods are refined and more sequences are added to DNA databases.

Conclusions

In summary, the paleoecological record offers a wealth of information to restoration ecologists. Multiple sources of information may be utilized to develop the history of vegetation occupying a site, which may provide a clearer picture of restoration reference conditions. More importantly, if there is enough temporal resolution in the paleoecological record some indication of variation in site vegetation may give restorationists a sense of historical variability of vegetation at the site and allow this variation to be incorporated into restoration designs. Accounting for this variation may pave the way for restorations that increase ecosystem function and resiliency.

References

Abbe TB, Montgomery DR (1996) Large woody debris jams, channel hydraulics, and habitat formation in large rivers. Regul Rivers: Res Manage 12:201–221

Abrams MD, Nowacki GJ (2008) Native Americans as active and passive promoters of mast and fruit trees in the eastern USA. Holocene 18:1123–1137

Anderson B, Rutherfurd I, Western A (2006) An analysis of the influence of riparian vegetation on the propagation of flood waves. Environ Model Software 21:1290–1296

Baker RG, Drake P (1994) Holocene history of prairie in midwestern United States: pollen versus plant macrofossils. Ecoscience 1:333–339

Baker RG, Fredlund GG, Mandel RD, Bettis III EA (2000) Holocene environments of the central great plains: multi-proxy evidence from alluvial sequences, southeastern Nebraska. Quatern Int 67:75–88

Bakker JP, Poschlod P, Strijkstra RJ, Bekker RM, Thompson K (1996) Seed banks and seed dispersal: important topics for restoration ecology. Acta Bot Neerl 45:461–490

Barnekow L, Loader NJ, Hicks S, Froyd CA, Goslar T (2007) Strong correlation between summer temperature and pollen accumulation rates for *Pinus sylvestris, Picea abies* and *Betula* spp. in a high-resolution record from northern Sweden. J Quatern Sci 22:653–658

Beaudoin AB (2007) On the laboratory procedure for processing unconsolidated sediment samples to concentrate subfossil seed and other plant macroremains. J Paleolimnol 37:301–308

Bedford B (1999) Cumulative effects on wetland landscapes: links to wetland restoration in the United States and Canada. Wetlands 19:775–788

Behrensmeyer AK, Hook RW (1992) Paleoenvironmental contexts and taphonomic modes in the terrestrial fossil record. In: Behrensmeyer AK, Damuth J, DiMichele WA, Potts R, Sues H-D, Wing SL (eds) Terrestrial ecosystems through time. University of Chicago Press, Chicago, pp 15–138

Bennett KD, Hicks S (2005) Numerical analysis of surface and fossil pollen spectra from northern Fennoscandia. J Biogeogr 32:407–423

Bennett KD, Parducci L (2006) DNA from pollen: principles and potential. Holocene 16:1031–1034

Bennington JB, Dimichele WA, Badgley C, Bambach RK, Barrett PM, Behrensmeyer AK, Bobe R, Burnham RJ, Daeschler EB, Dam JV, Eronen JT, Erwin DH, Finnegan S, Holland SM, Hunt G, Jablonski D, Jackson ST, Jacobs BF, Kidwell SM, Koch PL, Kowalewski MJ, Labandeira CC, Looy CV, Lyons SK, Novack-Gottshall PM, Potts R, Roopnarine PD, Stromberg CA, Sues H, Wagner PJ, Wilf P, Wing SL (2009) Critical issues of scale in paleoecology. Palaios 24:1–4

Benthardt K, Koch M, Kropf M, Ulbel E, Webhofer J (2008) Comparison of two methods characterizing the seed bank of amphibious plants in submerged sediments. Aquat Bot 88:171–177

Berggren G (1969) Atlas of seeds and small fruits of northwest European plant species, part 2. *Cyperaceae.* Swedish Natural Science Research Council, Stockholm, p 68

Berglund BE (1986) Handbook of Holocene paleoecology and paleohydrology. Wiley, New York, p 869

Bilby RE, Bisson PA (1998) Function and distribution of large woody debris. In: Naiman RJ, Bilby BE, Kantor S (eds). River ecology and management: lessons from the pacific coastal ecoregion. Springer-Verlag, New York, pp 324–338

Birks HJB, Birks HH (1980) Quaternary paleoecology. Arnold, London, p 289

Black BA, Abrams MD (2001) Analysis of temporal variation and species-site relationships of witness tree data southeastern Pennsylvania. Can J Forest Res 31:419–429

Bowen DE, Simon MP, Davis JW, Cope TM, Cusumano ZT, Hellmer JC, Winder VL, Soard SJ, Lidolph AM, Zielinski SE, James B, Runchey M, Hackmann T (2004) A list of plants observed along the lower Missouri River by the Lewis and Clark expedition in 1804 and 1806. Trans Kansas Acad Sci 107:55–68

Brown AD (2010) Pollen analysis and planted ancient woodland restoration strategies: a case study from the Wentwood, southeast Wales, UK. Veg Hist Archaeobot 19:79–90

Brown KJ, Pasternack GB (2005) A paleoenvironmental reconstruction to aid in the restoration of floodplain and wetland habitat on an upper deltaic plain, California, USA. Environ Conserv 32:1–14

Burney DA, Burney LP (2007) Paleoecology and "inter-situ" restoration on Kaua'i, Hawai'i. Front Ecol Environ 5:483–490

Burnham RJ, Wing SL, Parker GG (1992) The reflection of deciduous forest communities in leaf litter: implications for autochthonous litter assemblages from the fossil record. Paleobiology 18:30–49

Cairns J, Heckman JR (1996) Restoration ecology: the state of an emerging field. Ann Rev Energy Environ 21:167–189

Calcotte R (1995) Pollen source area and pollen productivity: evidence from forest hollows. J Ecol 83:591–602

Calcotte R (1998) Identifying forest stand types using pollen from forest hollows. Holocene 8:423–432

Carmichael D (1980) A record of environmental change during recent millenia in the Hackensack tidal marsh, New Jersey. Bull Torrey Bot Club 107:514–524

Chase MW, Fay MF (2009) Barcoding of plants and fungi. Science 325:682–683

Clague JJ, Turner RJW, Reyes AV (2003) Record of recent river channel instability, Cheakamus Valley, British Columbia. Geomorphology 53:317–332

Clewell AF, Rieger JP (1997) What practitioners need from restoration ecologists. Restor Ecol 5:350–354

de Wet A, Williams CJ, Tomlinson J, Carlson Loy E (2011) Stream and sediment dynamics in response to Holocene landscape changes in Lancaster County, Pennsylvania. In: LePage BA (ed) Wetlands—integrating multidisciplinary concepts. Springer, Dordrecht

DiMichele WA, Gastaldo RA (2008) Plant paleoecology in deep time. Ann Mo Bot Gard 95:144–198

Egawa C, Koyama A, Tsuyuzaki S (2009) Relationships between the developments of seedbank, standing vegetation and litter in a post-mined peatland. Plant Ecol 203:217–228

Ehrenfeld JG (2000a) Defining the limits of restoration: the need for realistic goals. Restor Ecol 8:2–9

Ehrenfeld JG (2000b) Evaluating wetlands within an urban context. Ecol Eng 15:253–265

Ferguson DK, Hofmann CC, Denk T (1999) Taphonomy: field techniques in modern environments. In: Jones TP, Rowe NP (eds) Fossil plants and spores: modern techniques. Geological Society, London, pp 210–213

Ficetola GF, Miaud C, Pompanon F, Taberlet P (2008) Species detection using environmental DNA from water samples. Biol Lett 4:423–425

Finkeldey R, Leinemann L, Gailing O (2010) Molecular genetic tools to infer the origin of forest plants and wood. Appl Microbiol Biotechnol 85:1251–1258

Flood RJ (1986) Seed identification handbook. National Institute of Agricultural Botany, Cambridge, p 72

Gastaldo RA, Douglass DP, McCarroll SM (1987) Origin, characteristics and provenance of plant macrodetritus in a Holocene crevasse splay, Mobile Delta, Alabama. Palaios 2:229–240

Gastaldo RA, Bearce SC, Degges C, Hunt RJ, Peebles MW, Violette DL (1989) Biostratinomy of a Holocene oxbow lake: a backswamp to mid-channel transect. Rev Palaeobot Palynol 58:47–60

Gastaldo RA, Riegel W, Püttmann W, Linnemann UH, Zetter R (1998) A multidisciplinary approach to reconstruct the late Oligocene vegetation in central Europe. Rev Palaeobot Palynol 101:71–94

Grimm EC (1993) Tilia (Version 2.0.b.4) and Tilia Graph (Version 2.0.b.5). Illinois State Museum, Springfield

Gross KL (1990) A comparison of methods for estimating seed numbers in the soil. J Ecol 78:1079–1093

Gugerli F, Parducci L, Petit RJ (2005) Ancient plant DNA: review and prospects. New Phytol 166:409–418

Gurnell A, Tockner K, Edwards P, Petts G (2005) Effects of deposited wood on biocomplexity of river corridors. Front Ecol Environ 3:377–382

Gurnell AM, Piégay H, Swanson F, Gregory S (2002) Large wood and fluvial processes. Freshwater Biology 74:601–619

Gwin SE, Kentula ME, Shaffer PW (1999) Evaluating the effects of wetland regulation through hydrogeomorphic classification and landscape profiles. Wetlands 19:477–489

Heusser CJ (1949) A note on buried cedar logs at Secaucus. N J Bull Torrey Bot Club 76:305–306

Heusser CJ (1963) Pollen diagrams from three former cedar bogs in the Hackensack tidal marsh, northeastern New Jersey. Bull Torrey Bot Club 90:16–28

Hilderbrand RH, Watts AC, Randle AM (2005) The myths of restoration ecology. Ecol Soc 10:19

Hoadley RB (1990) Identifying wood: accurate results with simple tools. The Tauton Press, Newtown, p 240

Hobbs RJ, Harris JA (2001) Restoration ecology: repairing the Earth's ecosystems in the new millennium. Restor Ecol 9:239–246

Hobbs RJ, Norton DA (2004) Ecological filters, thresholds and gradients in resistance to ecosystem reassembly. In: Temperton V, Hobbs RJ, Halle RJ, Fattorini M (eds) Assembly rules and ecosystem restoration. Island Press, Washington, pp 72–95

Hofreiter M, Mead JI, Martin P, Poinar HN (2003) Molecular caving. Curr Biol 13:R693–R695

Hollingsworth PM, Forrest LL, Spouge JL, Hajibabaei M, Ratnasingham S, van der Bank M, Chase MW, Cowan RS, Erickson DL, Fazekas AJ, Graham SW, James KE, Kim K, Kress WJ, Schneider H, van AlphenStahl J, Barrett SC, van den Berg C, Bogarin D, Burgess KS, Cameron KM, Carine M, Chacón J, Clark A, Clarkson JJ, Conrad F, Devey DS, Ford CS, Hedderson TA, Hollingsworth ML, Husband BC, Kelly LJ, Kesanakurti PR, Kim JS, Kim Y, Lahaye R, Lee H, Long DG, Madriñán S, Maurin O, Meusnier I, Newmaster SG, Park C, Percy DM, Petersen G, Richardson JE, Salazar GA, Savolainen V, Seberg O, Wilkinson MJ, Yi D, Little DP (2009) A DNA barcode for land plants. Proc Natl Acad Sci U S A 106:12794–12797

Hopfensperger KN (2007) A review of similarity between seed bank and standing vegetation across ecosystems. Oikos 116:1438–1448

Hough RB (1957) Hough's encyclopaedia of American woods. R. Speller, New York

Houlahan J, Findlay CS (2004) Estimating the "critical" distance at which adjacent land-use degrades wetland water and sediment quality. Landscape Ecol 19:677–690

Hurd EG, Goodrich S, Shaw NL (1994) Field guide to intermountain rushes. General Technical Report INT-306. Intermontane Research Station. Forest Service, United States Department of Agriculture, Ogden, p 56

Hurd EG, Shaw NL, Mastrogiuseppe J, Smithman LC, Goodrich S (1998) Field guide to intermountain sedges. General Technical Report RMRS-GTR-10. Rocky Mountain Research Station. Forest Service, United States Department of Agriculture, Ogden, p 282

Hyatt TL, Naiman RJ (2001) The residence time of large woody debris in the Queets River, Washington. Ecol Appl 11:191–202

Jackson ST (1997) Documenting natural and human-caused plant invasions using paleoecological methods. In: Luken JO, Thieret JW (eds) Assessment and management of plant invasions. Springer Verlag, New York, pp 37–55

Jackson ST, Hobbs RJ (2009) Ecological restoration in the light of ecological history. Science 325:567–569

Jacobson GL, Bradshaw RH (1981) The selection of sites for paleoenvironmental studies. Quatern Res 16:80–96

Jones JB, Smock LA (1991) Transport and retention of particulate organic matter in two low-gradient headwater streams. J NABS 10:115–126

Jones TP, Rowe NP (1999) Fossil plants and spores. The Geological Society Publishing House, London, p 396

Juggins S (2007) C2 Version 1.5: software for ecological and palaeoecological data analysis and visualisation. University of Newcastle, Newcastle upon Tyne

Jutila HM (2003) Germination in Baltic coastal wetland meadows: similarities and differences between vegetation and seed bank. Plant Ecol 166:275–293

Keane RE, Hessburg PF, Landres PB, Swanson FJ (2009) The use of historical range and variability (HRV) in landscape management. For Ecol Manag 258:1025–1037

Kellogg R, Rowe S (1981) An anatomical method for differentiating woods of western and mountain hemlock, a research note. Wood Fiber Sci 13:166–168

Kooistra MJ, Kooistra LI, van Rijn P, Sass-Klaassen U (2006) Woodlands of the past—the excavation of wetland woods at Zwolle-Stadshagen (the Netherlands): reconstruction of the wetland wood in its environmental context. Neth J Geosci—Geol Mijnbouw 85:37–60

Kowalski K, Wilcox D (1999) Use of historical and geospatial data to guide the restoration of a Lake Erie coastal marsh. Wetlands 19:858–868

Laderman AD (2003) Why does the freshwater genus *Chamaecyparis* hug marine coasts? In: Atkinson RB, Belcher RT, Brown DA, Perry JE (eds) Atlantic white cedar restoration ecology and management: proceedings of a symposium. Christopher Newport University, Newport News, pp 1–30

Landres PB, Morgan P, Swanson FJ (1999) Overview of the use of natural variability concepts in managing ecological systems. Ecol Appl 9:1179–1188

Larson PR (1963) Stem development of forest trees. Forest Sci Monograph 5:1–42

Lavoie C, Zimmermann C, Pellerin S (2001) Peatland restoration in southern Québec (Canada): a paleoecological perspective. Ecoscience 8:247–258

LeBlanc DC (1990) Relationship between breast-height and whole-stem growth indices for red spruce on Whiteface Mountain, New York. Can J For Res 20:1399–1407

Leck MA, Leck CF (2005) Vascular plants of a Delaware River tidal freshwater wetland and adjacent terrestrial areas: seed bank and vegetation comparisons of reference and constructed marshes and annotated species list. J Torrey Bot Soc 132:323–354

Leck MA, Schütz W (2005) Regeneration of Cyperaceae, with particular reference to seed ecology and seed banks. Perspect Plant Ecol Evol Syst 7:95–133

Liepelt S, Sperisen C, Deguilloux M, Petit RJ, Kissling R, Spencer M, de Beaulieu J, Taberlet P, Gielly L, Ziegenhagen B (2006) Authenticated DNA from ancient wood remains. Ann Bot (Lond) 98:1107–1111

Magee TK, Ernst TL, Kentula ME, Dwire KA (1999) Floristic comparison of freshwater wetlands in an urbanizing environment. Wetlands 19:517–534

Magee TK, Kentula ME (2005) Response of wetland plant species to hydrologic conditions. Wetlands Ecol Manage 13:163–181

Magyari E, Sümegi P, Braun M, Jakab G, Molnár M (2001) Retarded wetland succession: anthropogenic and climatic signals in a Holocene peat bog profile from North-East Hungary. J Ecol 89:1019–1032

Marshall S (2004) The Meadowlands before the commission: three centuries of human use and alteration of the Newark and Hackensack Meadows. Urban Habitats 2:4–27

Martin AC, Barkley WD (1961) Seed identification manual. University of California Press, Berkeley, p 221

Mauchamp A (1997) Threats from alien plant species in the Galápagos Islands. Conserv Biol 11:260–263

Middleton BA (2003) Soil seed banks and the potential restoration of forested wetlands after farming. J Appl Ecol 40:1025–1034

Montgomery FH (1976) Seeds and fruits of plants of Eastern Canada and Northeastern United States. University of Toronto Press, Toronto, p 232

Mylecraine KA, Zimmermann GL, Kuser JE (2005) Performance of Atlantic White-Cedar plantings along water table gradients at two sites in the New Jersey Pinelands. In: Burke MK, Sheridan P (eds) Atlantic white cedar: ecology, restoration, and management. Proceedings of the Arlington Echo Symposium. United States Department of Agriculture, Forest Service, Southern Research Station, Asheville, pp 7–10

National Research Council (1992) Restoration of aquatic ecosystems: science, technology, and public policy. The National Academies Press, Washington, p 576

Neill C, Bezerra MO, McHorney R, O'Dea CB (2009) Distribution, species composition and management implications of seed banks in southern New England coastal plain ponds. Biol Conserv 142:1350–1361

Palmer MA, RF Ambrose, Poff NL (1997) Ecological theory and community restoration ecology. Restor Ecol 5:291–300

Panshin AJ, deZeeuw C (1980) Textbook of wood technology: structure, identification, properties, and uses of the commercial woods of the United States and Canada, 4th edn. McGraw-Hill Book Co., New York, p 736

Parducci L, Suyama Y, Lascoux M, Bennett K (2005) Ancient DNA from pollen: a genetic record of population history in Scots pine. Mol Ecol 14:2873–2882

Payette S, Delwaide A (2004) Dynamics of subarctic wetland forests over the past 1500 years. Ecol Monographs 74:373–391

Pederson DC, Peteet DM, Kurdyla D, Guilderson T (2005) Medieval warming, little ice age, and european impact on the environment during the last millennium in the lower Hudson Valley, New York, USA. Quatern Res 63:238–249

Peglar SM (1993) The mid-Holocene Ulmus decline at Diss Mere, Norfolk, UK: A year-by-year pollen stratigraphy from annual laminations. Holocene 3:1–13

Pellerin S, Lavoie C (2003) Reconstructing the recent dynamics of mires using a multitechnique approach. J Ecol 91:1008–1021

Peterson JE, Baldwin AH (2004) Variation in wetland seed banks across a tidal freshwater landscape. Am J Bot 91:1251–1259

Poiani K, Johnson WC (1988) Evaluation of the emergence method in estimating seed bank composition of prairie wetlands. Aquat Bot 32:91–97

Pregitzer KS, Reed DD, Bornhorst TJ, Foster DR, Mroz GD, McLachlan JS, Laks PE, Stokke DD, Martin PE, Brown SE (2000) A buried spruce forest provides evidence at the stand and landscape scale for the effects of environment on vegetation at the Pleistocene/Holocene boundary. J Ecol 88:45–53

Price JN, Wright BR, Gross CL, Whalley WRDB (2010) Comparison of seedling emergence and seed extraction techniques for estimating the composition of soil seed banks. Methods Ecol Evol. doi:10.1111/j.2041-210X.2010.00011.x

Rheinhardt RD, McKenney-Easterling M, Brinson MM, Masina-Rubbo J, Brooks RP, Whigham DF, O'Brien D, Hite JT, Armstrong BK (2009) Canopy composition and forest structure provide restoration targets for low-order riparian ecosystems. Restor Ecol 17:51–59

Rich F (1989) A review of the taphonomy of plant remains in lacustrine sediments. Rev Palaeobot Palynol 58:33–46

Roberts HA (1981) Seed banks in soils. Adv Appl Biol 6:1–55

Ruhlman MB, Nutter WL (1999) Channel morphology evolution and overbank flow in the Georgia Piedmont. J Am Water Resour Assoc 35:277–290

Schauffler M, Jacobson GL Jr (2002) Persistence of coastal spruce refugia during the Holocene in northern New England, USA, detected by stand-scale pollen stratigraphies. J Ecol 90:235–250

Schlumbaum A, Tensen M, Jaenicke-Despres V (2008) Ancient plant DNA in archaeobotany. Veg Hist Archaeobot 17:233–244

Schopmeyer CS (1974) Seeds of woody plants in the United States, Agriculture Handbook 450. United States Department of Agriculture, Forest Service, Washington, p 883

Schweingruber FH (1990) Microscopic wood anatomy; structural variability of stems and twigs in recent and subfossil woods from Central Europe, 3rd edn. Eidgen–ssische Forschungsanstalt WSL, Birmensdorf, p 226

Shure DJ, Gottschalk MR, Parsons KA (1986) Litter decomposition processes in a floodplain forest. Am Midl Nat 115:314–327

Sipple W (1972) The past and present flora and vegetation of the Hackensack Meadows. Bartonia 4:4–25

Society for Ecological Restoration International Science and Policy Working Group (2004) SER International Primer on Ecological Restoration Society for Ecological Restoration International, Version 2. Society for Ecological Restoration International, Tucson, Arozona. www.ser.org

Starr JR, Naczi RFC, Chouinard BN (2009) Plant DNA barcodes and species resolution in sedges (Carex, Cyperaceae). Mol Ecol Resour 9:151–163

Sugita S (1994) Pollen representation of vegetation in quaternary sediments: theory and method in patchy vegetation. J Ecol 82:881–897

Suyama Y, Gunnarsson U, Parducci L (2008) Analysis of short DNA fragments from Holocene peatmoss samples. Holocene 18:1003–1006

Swanson FJ, Jones JA, Wallin DO, Cissel JH (1994) Natural variability—implications for ecosystem management. Volume II: ecosystem management principles and applications. In: Jensen ME, Bourgeron PS (eds) Eastside forest. Ecosystem health assessment. United States Department of Agriculture, Forest Service, Pacific Northwest Research Station, Portland, pp 80–94

ter Heerdt GNJ, Verwey GL, Bekker RM, Bakker JP (1996) An improved method for seed bank analysis: seedling-emergence after removing the soil by sieving. Funct Ecol 10:144–151

Tennessen D, Blanchette RA, Windes TC (2002) Differentiating aspen and cottonwood in prehistoric wood from the Chacoan great house ruins. J Archaeol Sci 29:521–527

Tiner RW, Swords JQ, McClain BJ (2002) Wetland status and trends for the Hackensack Meadowlands. An Assessment Report from the U.S. Fish and Wildlife Service's National Wetlands Inventory Program. United States Fish and Wildlife Service, Northeast Region, Hadley, Massachusetts, p 29

Ungar IA, Woodell SRJ (1996) Similarity of seed banks to aboveground vegetation in grazed and ungrazed salt marsh communities on the Gower Peninsula, South Wales. Int J Plant Sci 157:746–749

van der Putten N, Verbruggen C, Ochyra R, Spassov S, de Beaulieu J, De Dapper M, Hus J, Thouveny N (2009) Peat bank growth, Holocene palaeoecology and climate history of South Georgia (sub-Antarctica), based on a botanical macrofossil record. Quatern Sci Rev 28:65–79

van der Valk AG, Pederson RL, Davis CB (1992) Restoration and creation of freshwater wetlands using seed banks. Wetlands Ecol Manage 1:191–197

van der Valk AG, Bremholm TL, Gordon E (1999) The restoration of sedge meadows: seed viability, seed germination requirements, and seedling growth of Carex species. Wetlands 19:756–764

van Leeuwen JFN, Froyd CA, van der Knaap WO, Coffey EE, Tye A, Willis KJ (2008) Fossil pollen as a guide to conservation in the Galapagos. Science 322:1206

Vécrin MP, Diggelen RV, Grevilliot F, Muller S (2002) Restoration of species-rich flood-plain meadows from abandoned arable fields in NE France. Appl Veg Sci 5:263–270

Voli M, Merritts D, Walter R, Ohlson E, Datin K, Rahnis M, Kratz L, Deng W, Hilgartner W, Hartranft J (2009) Preliminary reconstruction of a pre-European settlement valley bottom wetland, southeastern, Pennsylvania. Water Resour Impact 11:11–13

Walker LR, Walker J, Moral RD (2007) Forging a new alliance between succession and restoration. In: Walker LR, Walker J, Hobbs RJ (eds) Linking restoration and ecological succession. Springer, New York, pp 1–18

Walker S, Lee WG, Rogers GM (2003) The woody vegetation of central Otago, New Zealand: its present and past distribution and future restoration needs. Sci Conserv 226:5–82

Walter RC, Merritts DJ (2008) Natural streams and the legacy of water-powered mills. Science 319:299–304

Wanner H, Beer J, Bütikofer J, Crowley TJ, Cubasch U, Flückiger J, Goosse H, Grosjean M, Joos F, Kaplan JO, Küttel M, Müller SA, Prentice IC, Solomina O, Stocker TF, Tarasov P, Wagner M, Widmann M (2008) Mid- to late Holocene climate change: an overview. Quatern Sci Rev 27:1791–1828

White PS, Walker JL (1997) Approximating nature's variation: selecting and using reference information in restoration ecology. Restor Ecol 5:338–349

Willerslev E, Hansen AJ, Binladen J, Brand TB, Gilbert MTP, Shapiro B, Bunce M, Wiuf C, Gilichinsky DA, Cooper A (2003) Diverse plant and animal genetic records from Holocene and Pleistocene sediments. Science 300:791–795

Willerslev E, Cappellini E, Boomsma W, Nielsen R, Hebsgaard MB, Brand TB, Hofreiter M, Bunce M, Poinar HN, Dahl-Jensen D, Johnsen S, Steffensen JP, Bennike O, Schwenninger J, Nathan R, Armitage S, de Hoog C, Alfimov V, Christl M, Beer J, Muscheler R, Barker J, Sharp M, Penkman KEH, Haile J, Taberlet P, Gilbert MTP, Casoli A, Campani E, Collins MJ

(2007) Ancient biomolecules from deep ice cores reveal a forested southern Greenland. Science 317:111–114

Williams CJ (2007) High-latitude forest structure: methodological considerations and insights on reconstructing high-latitude fossil forests. Bull Peabody Mus Nat Hist 48:339–357

Williams CJ, Johnson AH, LePage BA, Vann DR, Sweda T (2003) Reconstruction of Tertiary *Metasequoia* forests II. Structure, biomass and productivity of Eocene floodplain forests in the Canadian Arctic. Paleobiology 29:271–292

Williams CJ, Mendell EK, Murphy J, Court WM, Johnson AH, Richter SL (2008) Paleoenvironmental reconstruction of a middle Miocene forest from the western Canadian Arctic. Palaeogeogr Palaeoclimatol Palaeoecol 261:160–176

Wing SL, DiMichele WA, Phillips TL, Taggart R, Tiffney BH, Mazer SJ (1992) Ecological characterization of fossil plants. In: Behrensmeyer AK, Damuth JD, DiMichele WA, Potts R, Sues HD, Wing SL (eds) Terrestrial ecosystems through time: evolutionary paleoecology of terrestrial plants and animals. University of Chicago Press, Chicago, pp 139–180

Yamakawa C, Momohara A, Nunotani T, Matsumoto M, Watano Y (2008) Paleovegetation reconstruction of fossil forests dominated by *Metasequoia* and *Glyptostrobus* from the late Pliocene Kobiwako Group, central Japan. Paleontol Res 12:167–180

Yu ZC (2006) Holocene carbon accumulation of fen peatlands in boreal western Canada: complex ecosystem response to climate variation and disturbance. Ecosystems 9:1278–1288

Yu ZC, McAndrews JH, Siddiqi D (1996) Influences of Holocene climate and water levels on vegetation dynamics of a lakeside wetland. Can J Bot 74:1602–1615

Zedler JB (2000) Progress in wetland restoration ecology. Trends Ecol Evol 15:402–407

Part II
The Ecological Framework

Chapter 5
Classification of Wetlands

Mark M. Brinson

Abstract Humans have been classifying and managing wetlands for their ecological services and economic value that are specific to societal needs since the dawn of civilization. These classification systems and management strategies evolve as society's needs change both spatially and temporally. In general, most systems of classification can be divided into three broad categories that are based on their function (what wetlands do), structure (what wetlands look like), or utility (how wetlands are managed). Like any system of classification, it must be remembered that there is no one best classification approach and each is developed with a specific purpose. Among the structural systems, the National Wetland Inventory or Cowardin system is perhaps the most widely recognized. Functional systems are focused mainly on hydrology and external forces such as climate and include the hydrogeomorphic (HGM) approach. The utility approach is perhaps the most diverse among the systems of classification with the Millennium Ecosystem Assessment being the most recent iteration that is focused on the ecosystem services that provide some benefit to the individual as well as society. Here the three broad categories of wetland classification systems as well as their inter-relationships are discussed with the goal of recognizing the value(s) that each approach contributes for developing a more integrated and optimized approach for wetland management strategies.

Classification is an initial step in making sense of patterns in the natural and social sciences. The history of wetland classification is as long as that for other ecosystem types since wetlands are but one of a number of land cover types that includes uplands. In the early days of land classification, there was no overarching reason to draw undue attention to wetlands unless there was an emphasis on avoidance (they were "too wet to plow") or on their attraction (they were "good places to shoot ducks").

Published post-humously

M. M. Brinson (✉)
Biology Department, East Carolina University, Greenville, NC 27858, USA

B. A. LePage (ed.), *Wetlands,*
DOI 10.1007/978-94-007-0551-7_5, © Springer Science+Business Media B.V. 2011

From an ecological perspective, wetlands are simply a particular slice of the continuum between the wettest ecosystem types (rivers, lakes, estuaries, and the marine environment) and the driest environments within a physiographic (broad-scale subdivisions of the Earth's surface based on terrain texture, rock type, geologic structure, and history) region. From this perspective, one could take two approaches: either classify wetlands as part of an ecological gradient or recognize them as a distinct community type. This same dichotomy was the basis for one of the most controversial debates in ecology—that between Henry Allan Gleason, who argued for the so-called individualistic concept of the plant community and Frederick Edward Clements who countered with the community unit approach (eloquently described by Hagen 1992). R. H. Whittaker (1956) resolved the issue to some extent by pointing out that while plants tend to distribute themselves along an environmental continuum, it is the steepness of the gradient (wet to dry in the case of wetlands) that dictates how striking the boundary will be.

In more recent decades in the United States, the Clean Water Act, and the fall-out from recognizing "waters of the United States," created the need to draw a bright line between "waters" and the rest of the landscape. From a practical, and thus a social/management/political perspective, it became crucial to be able to draw boundaries and recognize that ecosystem types (or communities) actually exist as definable entities as Clements would have argued.

It turned out that the classification of wetlands as being distinct from uplands was not a trivial exercise, not just because the transition can be gentle and broad, possibly leading to ambiguous boundaries, but also because of the high stakes in the United States of regulating activities in wetlands on mainly private property (Kusler 1992). The issue was largely resolved in the mid 1990s (National Research Council 1995). For simplicity, I assume that the many definitions of wetlands have been adequately resolved elsewhere so that this chapter can focus on classifying wetland ecosystem types for which scientific consensus has been reached.

The history of wetland classification will only be briefly treated since it also has been adequately covered elsewhere[1] (National Council for Air and Stream Improvement, Inc 1991; Mitsch and Gosselink 2007). Here, the types of classification will be the focus with the realization that (1) we are getting better at classification as the knowledge base on wetlands improves and (2) we are becoming more sophisticated and strategic about how to classify wetlands for different purposes. In the end, there is no one "best" classification approach given the observation that the "best classification is the one that best serves the objectives for which it is developed."

---

[1] Steve Mader reviewed the existing literature for the NCASI (1991), which provides a rather extensive list of criteria that have been used to classify wetlands. They include water chemistry, hydrology, soils, vegetation, ecosystem and habitat type, physiographic and geomorphic, and waterfowl.

Types of Classifications

For simplicity and clarity, it is useful to recognize three approaches to classification that are consistent with the way that wetlands, and ecosystems in general, are commonly classified: *their structure, their functioning, and their utility*. The first of these is a focus on "how they appear," and can be called the "structural" approach. Keddy's (2000) classification comes up with four principal wetland types: swamp, marsh, bog, and fen, as well as two others that support truly aquatic plants (i.e., shallow water that is at least 25 cm in depth and wet meadow). Almost anyone can tell the difference between swamp, marsh, bog, and fen simply by looking at them. Structural classifications are particularly effective as the basis for mapping and inventory. As such, they serve one of the most fundamental sets of questions in natural resources management: how many kinds of wetlands are there, where are they, and how much is there of each type?

The second type of classification focuses on "what they do." We refer to this as the "functional" approach. It focuses on the differences between the way that wetlands receive and transport water, which is related to their shape and position in the landscape, recognizing that hydrologic regime is the most critical component of wetland behavior. This type of classification is particularly useful in controlling for natural variation as a prelude to assessment of ecological condition. It provides the foundation for insight into management alternatives, including the effects of negative impacts.

The third type of classification relates to "how they are managed" by the most powerful animal community on the planet, human society. We can think of this as the "utility" approach to classification. Early in the history of wetland management, a major goal was to convert them to uplands by draining or filling. We are now increasingly focused on the how wetlands are managed to provide human life support processes known as ecosystem services (Nyman 2011).

Of course the structural, functional, and utility approaches are interconnected and are becoming interdependent as we gain understanding of how to manage wetlands. In discussing these approaches in the pages that follow, the difficulty of separating them becomes more apparent as wetland management becomes more comprehensive and integrated. In fact, the outcome of this chapter is to recognize the value of all approaches, and to pose questions that, when considered, may lead to more optimal and coordinated approaches to wetland management.

Structural Approaches to Classification

One of the first steps in natural resources management is to locate and quantify the resource. For wetlands, this constitutes inventory and mapping. And in order to "inventory and map," classification is crucial.

The history of wetland nomenclature, and thus classification, goes back centuries, likely to the beginning of language development. Terms used in North America were most influenced by the northern European regions. In those areas peatlands were, and still are, very much a part of this recently glaciated landscape and thus were woven into the cultures of the region. As a result of the diversity of languages in this vast region, the names for wetland types were likewise diverse. We have such remarkable terms such as muskeg (Cree), moor (English), wad (Dutch), and Schwingmoor (German) that enrich our vocabulary to this day (not to mention flarks, paalsa bog, and aapa peatlands!).

During the past century, three national classification systems in the United States were developed prior to the National Wetland Inventory (NWI) approach of the United States Fish and Wildlife Service (USFWS) (Cowardin et al. 1979). The first in 1906 was, not ironically at the time, an attempt to determine the agricultural potential for remaining wetlands, authorized by the United States Department of Agriculture (USDA) (Stegman 1976 cited in National Council for Air and Stream Improvement, Inc 1991). The questionnaire for developing this database was distributed to persons in all counties, but excluded eight western states dominated by public lands and all tidally-influenced wetlands. The second inventory in 1922 had similar motivations of evaluating the potential for "reclamation." It was assembled from a variety of data sources including soil surveys and topographic maps as well as a 1920 census of drainage (National Council for Air and Stream Improvement, Inc 1991). The third national inventory broke rank by orienting its classification toward their importance for waterfowl and other wildlife and the emerging role of the USFWS in managing these resources. Commonly known as "Circular 39" (Shaw and Fredine 1956), the report presented a geographically stratified approach with the following categories: (1) inland fresh areas; (2) inland saline areas; (3) coastal fresh areas; and (4) coastal saline areas. For example, the coastal saline areas included salt flats, salt meadows, irregularly and regularly flooded salt marshes, sounds and bays, and mangrove swamps. For nearly two decades this classification was routinely used. It remained the principal basis for the development of wetland policy, regulation, and preservation until the advent of the NWI program of the USFWS (Cowardin et al. 1979).

The NWI is a truly hierarchical classification in large part to accommodate NWI mapping efforts. Often referred to as "Cowardin et al.", the classification included also deepwater habitats distinguished as "permanently flooded lands lying below the deepwater boundary of wetlands," or 2 m depth. Partly because the classification is the foundation for NWI mapping and inventory, it finds broad acceptance and use among wetland scientists. If one can point to a universal language among wetland practitioners in the United States, then it is the use of the coding system (e.g., PFO1A, E2EM1P6, etc.). The hierarchy steps down as follows: system; subsystem; class; dominance type; and finally modifiers that include water regime, water chemistry, and soil, as well as human-modified categories.

Of course other classifications are prominent and useful in North America. Principal among these is the Canadian system (Warner and Rubec 1997). It offers five classes with a number of forms and subforms for each class. The approach uses vegetation to some degree for separation. It recognizes swamp (forested) as being distinct from marsh (herbaceous), but also identifies shallow water and fens and bogs as classes that would support several types of vegetation cover. The advantage of this system of classification is that common wetland descriptors are used at the form and subform level, which leaves little room for ambiguity (e.g., blanket bog, floating fen, tidal freshwater swamp, tidal marsh, estuarine channel water).

In spite of the overarching acceptance and use of the NWI classification in the United States, a number of other structural classifications have been developed for specific purposes. Among these are regional and state-wide efforts to address particular wetland types or conform to state regulations and policies. A noteworthy example of a regional approach is classification of the prairie pothole wetlands (Kantrud et al. 1989), one of a number of community profiles produced by the Office of Biological Services of the USFWS. While these publications did not all include a system of classification, the recognition of "communities" in and of itself is an implicit classification decision. The community profiles, however, were not oriented toward inventory and mapping efforts. Rather, they synthesized existing information on the structure and functioning of wetland types to assist regulatory programs of the United States. In this sense they served more of a functional role (discussed next) than a purely structural purpose.

Most countries are signatories to the Ramsar Convention on Wetlands of International Importance (Gardner and Davidson 2011), an organization that has a distinct approach to classifying wetlands. To some extent, efforts of this organization in the United States have attracted little attention in comparison with Federal and State programs that regulate activities in wetlands. But its importance is without equal in many countries that do not have strong programs for managing wetlands. For example, MedWet, which was founded in 1991, encourages international collaboration among Mediterranean countries to protect wetlands (http://www.medwet.org/medwetnew/en/02.ABOUT/02.0.about_medwet.html). Designating a Ramsar Convention wetland requires a political commitment by a country at some of the highest levels of government. While the effectiveness of designation in terms of protecting wetlands is far from absolute, neither is any program.

Some of these structural approaches of course rely on the appearance of wetlands from the standpoint of vegetation; swamps have trees, coastal marshes don't. Others rely on landform shape (raised bogs, riparian zone). The problem with relying on plant species is that they respond not so much to landform shape as they do to climate, geographic location, local conditions of wetness, nutrient supply, and disturbance such as fire. Consequently, for the purposes of generalization and consistency, it would be attractive to have a "pure" structural classification that relies only secondarily on vegetation.

Such a structural classification was introduced by Semeniuk and Semeniuk (1995) in a special issue of *Vegetatio* (now *Plant Ecology*) that covered wetland classification in a variety of geographic regions of the world (Finlayson and van der Valk 1995). This system of classification combines landform setting with various types of hydroperiod resulting in 13 primary wetland types (Table 5.1). Like the NWI classification, lakes and rivers are included. The classification is non-genetic in that it does not recognize or rely on the origin of a wetland. A depression created

Table 5.1 Thirteen wetland categories determined by landform and water longevity. (From Semeniuk and Semeniuk [1995])

Water longevity	Landform				
	Basin	Channel	Flat	Slope	Highland
Permanent inundation	Lake	River	–	–	–
Seasonal inundation	Sumpland	Creek	Floodplain	–	–
Intermittent inundation	Playa	Wadi	Barkarra	–	–
Seasonal waterlogging	Dampland	Trough	Palusplain	Paluslop	Palusmont

by glaciation would not be distinguished from one resulting from wind deflation, all other factors being equal.

A problem with many structural classifications is that they use mixed criteria, most generally vegetation and geomorphology. Using vegetation alone has shortcomings because the same plant species commonly inhabit different habitat types, or geomorphic (landform) settings. The mixing of criteria is especially true of the Ramsar Convention on Wetlands of International Importance listing of wetland types. Semenuik and Semeniuk (1997) applied their approach to the first tier of 24 inland types of the Ramsar classification. This "cross walk" between the two systems of classification was intended to reconcile the two systems by revealing the inherent underlying geomorphic structure of the Ramsar list and its hierarchical framework. The paper also criticizes the internally diverse class "palustrine" in the NWI system, a term that is synonymous with wetland.

Functional Approaches to Classification and Landscape Diversity

Functional classification systems focus more on hydrologic regime and other external forces including climate. One of the most intriguing classification systems was developed decades ago for coastal ecosystems including wetlands by Odum et al. (1974). It recognized the role of seasonality and stressors as major influences of ecosystem functioning. Ecosystem types included "naturally stressed systems with wide latitudinal range" (i.e., high energy beaches, rocky intertidal rocks), "natural tropical ecosystems of high diversity" (i.e., mangrove swamps, coral reefs), and even "emerging new systems associated with man" (i.e., systems receiving sewage waste, thermal pollution, pulp mill discharge). While this system is not in use today, it was arguably the inspiration for other functionally-based efforts.

One example of this approach was developed for mangrove swamps (Lugo and Snedaker 1974). It incorporates components of both geomorphology (the study of landforms and the processes that shape them) and vegetation structure. The types of mangrove swamps include fringe, riverine, basin, overwash, and dwarf. The last of these includes red mangroves growing on what were suspected to be nutrient poor soils in South Florida, USA. More recent work demonstrated that these dwarf mangroves are stunted by phosphorus-deficiency, rather than by high salinity (Medina et al. 2009).

Another approach, limited to inland marshes, emphasized hydrologic gradients (Gosselink and Turner 1978). It recognized water inputs, types of flow, as well as "hydro-pulse" responsible for partitioning the following types: raised-convex, meadow, sunken-convex, lotic, tidal, and lentic (Table 5.2). Each of the foregoing approaches recognizes that vegetation and other habitat features are the result, not the origin, of the forces that fashion wetlands.

In attempt to capture both tidal and inland wetlands, and to rely upon first principles (as much as possible for an ecological approach), the hydrogeomorphic (HGM) approach (Brinson 1993) built upon the seasonality and stressor concept of Odum et al. (1974) and the mangrove classification of Lugo and Snedaker (1974). The six classes are riverine, depressional, fringe, slope, lacustrine, and flats (Table 5.3). Two very distinct flat groups are organic soil and mineral soil flats, described in Brinson (2009). One could add geothermal wetlands to this list as yet another HGM type, although Semeniuk and Semeniuk (1997) suggest the contrary.

The HGM approach was developed principally for the purpose of partitioning wetlands into functional types as a prelude to a reference-based functional assessment. Underlying the HGM classification are the dominant and interacting factors of geomorphic setting, water source, and hydrodynamics on ecosystem functioning (Brinson 1993). The logic was that wetland ecosystems have optimized their functioning (nutrient cycling, energy flows, community maintenance, etc.) through millennia of selective processes that are consistent with the maximum power principle (Odum and Pinkerton 1955). Consequently, impacts, especially those that alter sources of water and hydrodynamics, will likewise negatively affect ecosystem functioning and result in degradation, and ultimately a change in state. In extreme cases wetlands would shift to non-wetlands. In practice, the classification is intended to partition the natural variation among wetland types so that variation due to human activities (i.e., alterations or impacts) will be more apparent, and easier to detect.

The HGM classification has gained acceptance as a reference-based approach (Brinson and Rheinhardt 1996) in its application to assessing the effects of restoration projects (Gwin et al. 1999), arraying the variation in lacustrine fringe wetlands in the Laurentian Great Lakes (Detenbeck et al. 1999), and in providing an organizing theme for an impact assessment framework in the United States (Smith et al. 1995; http://el.erdc.usace.army.mil/wetlands/wlpubs.html). In contrast, its usefulness is inadequate when applied to assessing certain types of detailed processes, such as nitrogen cycling (Ehrenfeld 2000; Standler and Ehrenfeld 2009) and hydrologic flow paths (Cole and Brooks 2000; Cole 2006). Another limitation derives from its dependence on relatively unaltered conditions of reference standard wetlands as a basis for comparison. Landscapes that have been settled and modified for centuries still have wetlands, but they are often fragmented remnants of much more extensive coverage. Yet they continue to function in ways that are considered to have ecological value by providing waterfowl habitat, denitrification, flood ame-

Table 5.2 Major hydrodynamic characteristics of marsh types. (Modified from Gosselink and Turner [1978])

Marsh type	Water inputs				Type of water flow			Perco-lation	Outputs		Hydropulse
	Capillarity	Precipita-tion	Upstream	Down-stream	Sub-surface	Surface sheet flow	Overbank flooding		Evapo-transpira-tion	Down-stream runoff	
Raised-convex	+	+			+			+	+		Seasonal
Meadow	+	+	*		+			+	+	*	Seasonal
Sunken-convex	+	+	+		+	**			+		Seasonal
Lotic	+	+	+		+	+	+		+	+	Seasonal
Tidal	+	+	+	+	+	+	+		+	+	Tidal
Lentic	+	+	++		+	+	+		+	+	Interannual

* Very little
** Very slow

Table 5.3 Example of the HGM classes, regional subclasses, and community types for wetlands of the Ozark Mountains region of Arkansas, US. (From Klimas et al. [2008])

Class	Regional subclass	Community type
Flat	Non-alkali flat	Hardwood flat
		Wet tallgrass prairie
Riverine	High gradient riverine	High-gradient riparian zone
	Mid-gradient riverine	Mid-gradient floodplain
		Mid-gradient backwater
	Low-gradient riverine	Low-gradient overbank
		Low-gradient backwater
	Impounded riverine	Beaver complex
	Spring run	Spring run
Depression	Connected depression	Floodplain depression
	Unconnected depression	Mountaintop depression
		Sinkhole
Fringe	Connected lacustrine fringe	Lake margin/shore
	Reservoir fringe	Reservoir shore
	Unconnected lacustrine fringe	Unconnected lake margin
Slope	Non-calcareous slope	Wet-weather seep
		Sandstone glade
	Calcareous slope	Calcareous perennial seep

lioration, and other ecosystem services. To address this, the European system has adopted hydrogeomorphic units (HGMUs) as the unit for classification and assessment (Maltby et al. 2009).

The factors considered in either the HGM or the HGMU classifications are not limited to land within wetland boundaries. Geomorphic setting indicates that a wetland "is set" in a larger region, water source implies that water is coming from elsewhere, and hydrodynamics within a wetland is often a result of conditions upstream from the wetland or climatic seasonality (or lack of it). The question to ask is—At what scale do meaningful patterns of wetlands begin that would usefully contribute to a classification system? This can be done by considering an array of spatial scales and thus by looking beyond a single wetland to evaluate the diversity of wetland types that a landscape supports.

A landscape approach to classifying wetlands can take a spatially explicit approach ranging from large intercontinental scales to some subunit of a particular wetland type. A nested hierarchical array would emerge as follows: global; continental; life zone or physiographic region; hydrogeomorphic complex (Bedford 1996); geomorphic setting (i.e., HGM); and hydrogeomorphic unit (i.e., HGMU). At each of these levels it is theoretically possible to compare the variation and diversity in wetland types within a particular level. The global scale is unique because we have no other known planet with wetlands for comparison. Likewise, planet Earth has too few continents to afford any degree of replication for analyzing patterns. For example, the comparison of wetlands between Australia with North America is unlikely to provide insight into wetland classification. At the life zone or physiographic scale, however, patterns begin to emerge, in part because climate is

Hydrogeologic landscape of a physiographic region

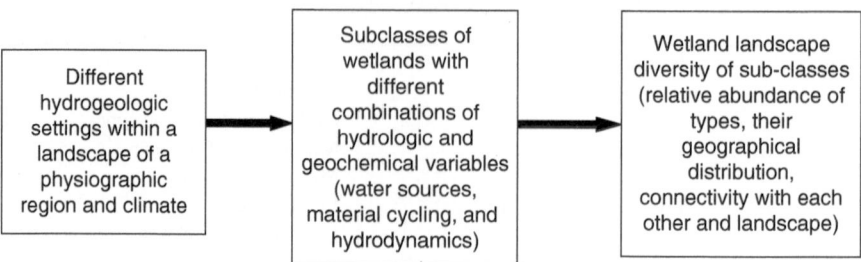

Fig. 5.1 Relationship of climatic and hydrogeologic setting to landscape diversity of wetland types. (Adapted from Bedford [1996])

homogeneous by definition and landform has repeated patterns that exert control on groups of wetlands of distinctive types. At this spatial scale, the relative abundance of different wetland types repeats itself.

Bedford (1996) argued that the physiographic region is where the occurrence of wetlands is highly predictable and is responsible for "wetland landscape diversity" (Fig. 5.1). Depressions will contain depressional wetlands driven by groundwater discharge and surface flows. In contrast, valleys will contain riverine wetlands and so forth. The predictability is built into the landscape unit and is dictated by how water is routed and detained. One of Bedford's points is that the diversity of settings within a particular landscape has implications for wetland restoration. She suggested that the landscape both constrains the abundance of wetlands and provides opportunities for wetland type diversity. Restoration efforts should attempt to adapt to this pattern and not "over-engineer" by restoring wetlands in the wrong place or restoring wetlands that will not be sustained. Gwin et al. (1999) report a now-classic example of the consequences of not replicating landscape diversity of wetlands in restoration programs.

This pattern of differences among physiographic regions was demonstrated by Winter (1992) for the United States. There, he identified eight settings; terraces and scarps within coastal wetlands; terraces within riverine valleys; steep slopes adjacent to narrow lowlands in mountains; large extensive lowland depressions (playas); morainal depressions; dune fields; sinkholes and other depressions; and permafrost. He aggregated these into climatic regions of cold, warm, wet, and dry. The combination of these extremes with the eight settings yields of total of 24 "type localities." Winter presented this matrix with the suggestion that a comparative approach for processes common to all wetlands may reveal patterns that are "unique to particular landscapes and climates."

The patterns described would be useful in setting up large research programs for wetlands at continental scales as Winter (1992) recommends. From an applied perspective, a practitioner working in one of the 24 type localities might quickly master knowledge about the relatively narrow set of conditions within the region. On the other hand, practitioners who work across a wider range of physiographic and climatic settings need to be aware of the fundamental differences among regions. In almost any physiographic region, except the coastal regions where wetlands are constrained to being tidal, and, of course excluding Antarctica, one encounters nearly all wetland types listed above, except in arid climates where wet flats cannot be supported because the principal source of water is precipitation.

Physiographic differences are also the basis for recognizing "regional subclasses" in the HGM approach. While the riverine wetland class occurs on every continent, the variation between the Paraná River, in Buenos Aires Province, Argentina, and Little Contentnea Creek, in North Carolina, USA, is hardly worth comparison because they differ so much in size. Recognition of the regional subclasses, that further classify and partition natural variation within a class, is critical for functional assessment and other purposes that rely on classification (Brinson and Rheinhardt 1996). For the HGM approach, the area that encompasses a regional subclass is known as the reference domain (Smith et al. 1995).

Utility Approaches to Classification: Managed Wetlands

To be fair, it should be noted that the classification systems just discussed were inspired by a motivation to manage wetlands for agriculture through drainage, as for the early United States classification systems, or for particular wildlife species, for which Circular 39 was assembled. For mapping and inventory, as described earlier, the structural approach would seem appropriate. For assessment of condition, a functional classification is useful. What, then, would be appropriate for a utility approach?

The "ecosystem services" concept described in the Millennium Ecosystem Assessment (2005) (http://www.millenniumassessment.org) provides a framework for a utility approach to classification. The "classes" of ecosystem services are *provisioning* (food, fresh water, fuel, wood, and fiber), *regulating* (climate, floods, disease regulation, and water purification), *cultural* (spiritual, aesthetic, educational, and recreational), and *supporting* (primary production, nutrient cycling, and soil formation). In a profound way, these four categories are the ultimate in "utility" for society in the sense that they are oriented directly toward human well-being whether or not it is explicitly acknowledged. Hereafter, only the provisioning and regulating components will be the focus of classification, although the consequence of management also affects supporting ecosystem services.

The Millennium Ecosystem Assessment framework goes well beyond the needs of individuals locally, such as food and fuel, and even encompasses global processes upon which societies depend for life support. Of course, wetlands have long been identified as providing a number of ecosystem services, not the least of which is carbon sequestration (Schlesinger 1977), as well as effects on methane and CO_2 emissions, all related to climate regulation. Feedbacks at the global scale are widely recognized. For example, greenhouse gas emissions from thawing permafrost peatlands are expected to place additional stress on distant coastal wetlands through accelerated rates of rising sea level. Thus, it is important to evaluate such effects, not from the strictly structural or functional approaches discussed earlier in this chapter, but to categorize wetlands according to their ecosystem services.

By first classifying the services and understanding their nature, management actions are better able to deal with the consequences of their interaction. The classification and interaction of ecosystem services is explained by Rodríguez et al. (2005). Ecosystem services have a spatial scale that ranges from local to global in effect, they have a temporal scale that indicates whether their effects take place rapidly or slowly, and they have an irreversibility property that describes the likelihood that a given perturbation, once terminated, will allow the system to return to its original state. This framework provides policy-makers with the tools to evaluate how far-reaching a particular set of choices will be (spatially and temporally) and whether it is reasonable to expect a return to the original state if a policy is implemented for this to take place. In this way, scenarios can be constructed and analyzed for both quantitative and qualitative effects. Some combinations of service choices may result in synergistic negative consequences that would be unanticipated if only one set of the service options was considered and evaluated. The challenge for policy-makers is to evaluate trade-offs among services for different scenarios such that negative impacts are minimized and the system can sustain services into the future. Within the confines of wetland management, this seems to be consistent with the recommendation of supply-side sustainability of Allen et al. (2002) in which eco-systems are managed "for the system that produces outputs rather than the outputs themselves."

Wetlands are managed toward different utility endpoints, ranging from protecting and maintaining their natural condition to converting them to landuses that jeopardize wetland functioning altogether. We will not consider ecosystem services of the latter case since conversion to upland landuses is beyond the scope of wetland classification. However, short of this point, wetlands have been and will continue to be managed or altered to different degrees. Table 5.4 provides examples of utility class endpoints oriented toward the provisioning of specific ecosystem services. The consequences of managing toward these utility endpoints are expressed as trade-offs that interfere with competing services in the provisioning, regulating, and supporting categories.

The utility class examples are ranked roughly in order of increasing intensification of consequences from the top to the bottom of the Table 5.4. As a general feature, utility classes that are managed for harvestable products, but with minor effects on the hydrologic regime, are considered to have the least effect on other ecosystem services. Waterfowl and furbearer harvests are relatively benign compared to the other examples provided in Table 5.4 because they do not alter hydropattern and biogeochemical conditions in the long run. Moderate nutrient loading of riparian wetlands may interfere little with other services, particularly if nitrate removal is the target, and the denitrification process is the focus of management for protecting downstream water quality (Peterjohn and Correll 1984).

Timber harvesting of bottomland hardwoods, if judiciously carried out, allows a forest to return to its original species composition within decades. On the other

Table 5.4 Examples of regional subclasses managed toward various utility class endpoints. The utility class column refers to provisioning and regulating services that are optimized or maximized through management toward a utility endpoint. The consequences column includes provisioning and regulating services that are co-opted or diminished as a result of management. Consequences of management generally intensify from top to bottom of the table as trade-offs of services become more extreme

Utility class endpoints (provisioning (P) or regulating (R) services)	Wetland type example	Consequences of management (examples of trade-offs with provisioning and regulating services)	Source
Waterfowl and furbearer harvest (P)	Prairie pothole	Altered community and food web structure	Weller (1988)
Nutrient and sediment removal from agriculture (R)	Forested riparian zone	Unresolved; minor if loading is low	Peterjohn and Correll (1984)
Waterfowl production enhanced with greentree reservoir (P)	Bottomland hardwood floodplain	Reduced tree growth and a shift in canopy species composition	King (1995)
Timber harvest after natural regeneration (P)	Bottomland hardwood floodplain	Temporarily reduced primary production and habitat quality during forest maturation	Kellison et al. (1998)
Hydraulic loading for stormwater storage (R)	Palustrine in urban settings	Amphibian habitat degradation due to greater water level fluctuations	Richter and Azous (1995)
Conversion to short rotation silviculture production of pine (P)	Wet hardwood flat	Reduced carbon sequestering; reduced plant diversity and altered community structure, especially for neotropical passerines	Dickson et al. (1995)
Nutrient removal from wastewater (R)	Peat bog in northern Michigan, USA	Shift from bog species to cattail dominance due to phosphorous loading	Kadlec and Bevis (2009)
Conversion of peatland to impoundment for hydroelectric power (P)*	Experimental peatland, Ontario, Canada	Change from sink to source for atmospheric carbon; accumulated methyl mercury released into atmosphere	Kelly et al. (1997)
Impoundment for waterfowl enhancement (P)*	*Juncus* marshes at sea level, North Carolina, USA	Loss in geomorphic capacity to respond to rising sea level; state change to estuary	M. Brinson, personal observations
Water abstraction for irrigation (P)*	Doñana National Park, Spain	Reduced wetland area from groundwater abstraction	Muñoz-Reinoso (2001)
Dyke construction for silviculture (P)*	Paraná River Delta, Argentina	Water storage and sediment deposition; altered habitat	Kandus et al. (2009)
Drainage for agriculture and silviculture in peatlands, (P)*	Pocosin peatlands, North Carolina	Loss of atmospheric carbon sequestering, landscape diversity, and wide ranging wildlife	Richardson (1983)

* Indicates considerable conversion to either upland or aquatic habitat

hand, altering the habitat through increased, but moderate water retention, as for greentree reservoirs, actually causes a shift in forest species composition, perhaps more than harvesting would (King 1995). Short-rotation silviculture has both short- and long-term consequences, including a long-term sacrifice in habitat quality and carbon sequestration. The degree of trade-offs for nutrient and sediment removal depends not only on what type and how much loading is occurring, but also what type of wetland is being altered. However, nutrient removal from wastewater by ombrotrophic wetlands may not only quickly exceed the capacity of the system to retain nutrients (Kadlec and Bevis 2009), but can alter species composition, food webs, and ecological character, as in the Florida Everglades (Davis and Ogden 1994). The trade-off between nutrient retention and biodiversity has been documented for constructed wetlands (Hansson et al. 2005).

Wetlands can be managed toward either drier or wetter conditions: for the former, the endpoint can be a non-wetland and the total loss of wetland functioning. Toward wetter conditions, an aquatic ecosystem may develop that represents a true change in ecosystem state and replacement with an array of alternative ecosystem services consistent with lakes and reservoirs. Five of these utility endpoints at the bottom of Table 5.4 diminish the surface area of wetlands simply as the result of too little or too much water.

Classification by utility could find application as an extension of structural and functional classifications. The work has not been done to fully develop this framework. The following is presented to prompt thought, rather than to present an established paradigm. But as it turns out, managers and regulators routinely evaluate trade-offs between alternative management endpoints. This is often done for wetlands in a larger socioeconomic context through impact analyses that deal with balancing costs and benefits. One could argue that the utility classification has components of a combined ecosystem condition assessment and impact analysis.

The utility classification is distinct from a functional classification due to its emphasis on ecosystem services. Further, unlike functional classifications that merely recognize departure from reference standard conditions, utility classifications explicitly identify what provisioning and regulating services will be enhanced and which services will be compromised. Rather than simply illustrating that the utility is fiber production (i.e., a pine plantation) or is the provisioning of recreation (i.e., a duck hunting impoundment), utility is defined in the context of both gains and losses of services as in Table 5.4. This provides a more comprehensive framework for policy-making and management options than that provided by functional classification and condition assessment.

Toward a More Integrated and Comprehensive Approach

Each of the approaches to wetland classification has merit as judged by application to inventory and mapping programs, to functional or condition assessment methodologies, to national programs that deal with the complexity of physiographic re-

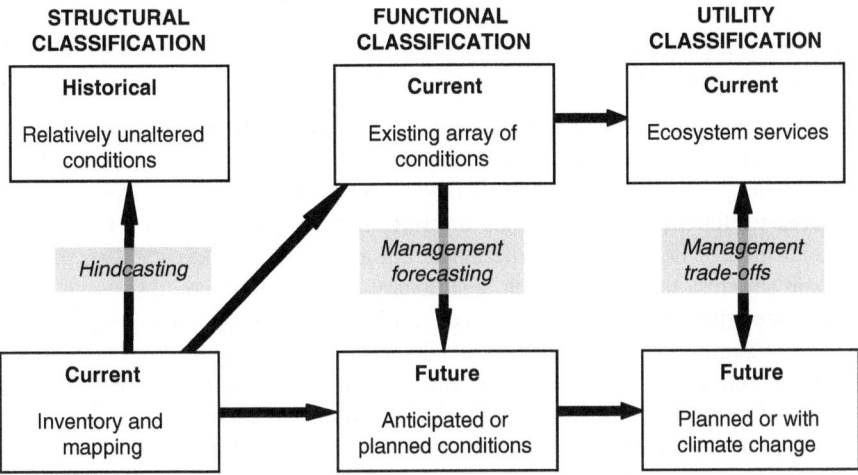

Fig. 5.2 Three classification approaches for wetlands discussed in this chapter. Each can be used for different purposes. The challenge is for wetland management programs to integrate them and to take advantage of the attributes of each

gions, and to the range of ecosystem services following the Millennium Ecosystem Assessment framework. However, the classifications described in the preceding sections tend to be used independently of one another. Historically, the structural approach preceded the functional, and the functional preceded the landscape. But utility knowledge about wetlands has been recognized from the time that the first wetland was drained or the first cranberry was eaten. Thus it is ironic that explicit classification by ecosystem services is a relatively recent phenomenon.

Wetland management has benefited by using attributes of each of the three classification frameworks (Fig. 5.2). For structural classification, inventory and mapping of current wetland distribution can be placed into an historical context by hindcasting the pre-Columbian situation. This may reveal opportunities toward which restoration efforts can be directed (de Wet et al. 2011; Williams 2011). For functional classification, current conditions can be evaluated and compared by forecasting anticipated or planned conditions. The effect of "building out" an urban area according to zoning regulations is one way to visualize how wetlands may be impacted by development. For utility classification, both current and future ecosystem services can be derived in part from the functional classification with one additional consideration—evaluation of climate change effects. Examples of management trade-offs between current and future ecosystem services are provided in Table 5.4. More complete analyses of examples such as these could be a valuable source of information for policy development. The evaluation of ecosystem service trade-offs must rely on underlying results of functional and condition assessments.

More effort is needed for wetland management programs to integrate the classification categories in useful ways. What follows are questions that may be worth considering in wetland programs regardless of their current status.

1. How might the various approaches to classification be brought together to help solve natural resource problems? Integration might be useful for existing, fragmented programs that have conducted wetland inventories, but lack a framework for functional classification or projecting ecosystem services. For regions with little or no inventory experience, but a substantial understanding of wetland structure through research, a structural classification would serve to allow extrapolation of site-specific research results to larger geographical regions that harbor similar kinds of wetlands.
2. What would be the most effective and efficient approach to combine or aggregate classification systems to optimize an approach for meeting a number of needs? The advantage of developing such a plan may eliminate duplication of effort and may render obsolete products that lack direct usefulness. By viewing classifications as one integrated framework for resource management, efficiencies of information development may be improved.
3. How might the classifications be integrated to inform programs about adaptation to climate change? Winter's (1992) landscape approach could be adapted to consider changes in wetland distribution and abundance with climate change by invoking a space-for-time approach introduced in Michener et al. (1997).

Classification is but one of many tools useful to wetland resource management and policy development. While it is recognized as one of the first steps in understanding patterns of natural complexity, classification can be more than simply descriptive. The three types of classification described in this paper represent increasing levels of understanding and awareness. The challenge is to build upon this framework in ways that lead to productive and sustainable management of wetland resources.

References

Allen TFH, Tainter JA, Hoekstra TW (2002) Supply-side sustainability. Columbia University Press, New York, p 459

Bedford BL (1996) The need to define hydrologic equivalence at the landscape scale for freshwater wetland mitigation. Ecol Appl 6:57–68

Brinson MM (1993) A hydrogeomorphic classification for wetlands. United States Army Corps of Engineers, Technical Report WRP-DE-4. Washington, p 101. http://el.erdc.usace.army.mil/wetlands/wlpubs.html

Brinson MM (2009) Chapter 22. The United States HGM (hydrogeomorphic) approach. In Maltby E, Barker T (eds) The wetlands handbook. Wiley-Blackwell, Oxford, pp 486–512

Brinson MM, Rheinhardt R (1996) The role of reference wetlands in functional assessment and mitigation. Ecol Appl 6:69–76

Cole CA (2006) HGM and wetland functional assessment: six degrees of separation from the data? Ecol Indicators 6:485–493

Cole CA, Brooks RP (2000) Patterns of wetland hydrology in the ridge and valley province, Pennsylvania, USA. Wetlands 20:438–447

Cowardin LM, Carter V, Golet FC, LaRoe ET (1979) Classification of wetlands and deepwater habitats of the United States. United States Fish and Wildlife Service, Washington, p 79

Davis SM, Ogden JC (1994) Everglades: the ecosystem and its restoration. St. Lucie Press, Delray Beach, p 826

de Wet A, Williams CJ, Tomlinson J, Carlson Loy E (2011) Stream and sediment dynamics in response to Holocene landscape changes in Lancaster County, Pennsylvania. In: LePage BA (ed) Wetlands—integrating multidisciplinary concepts. Springer, Dordrecht

Detenbeck NE, Galatowitsch SM, Atkinson J, Ball H (1999) Evaluating perturbations and developing restoration strategies for inland wetlands in the Great Lakes Basin. Wetlands 19:789–820

Dickson JG, Thompson III FR, Conner RN, Franzreb KE (1995) Silviculture in central and southeastern oak-pine forests. In: Martin TE, Finch DM (eds) Ecology and management of neotropical migratory birds. Oxford University Press, New York, pp 245–265

Ehrenfeld JG (2000) Evaluating wetlands within an urban context. Ecol Eng 15:253–265

Finlayson CM, van der Valk AG (1995) Wetland classification and inventory: a summary. Vegetatio 118:185–192

Gardner RC, Davidson NC (2011) The Ramsar convention. In: LePage BA (ed) Wetlands—integrating multidisciplinary concepts. Springer, Dordrecht

Gosselink JR, Turner RE (1978) The role of hydrology in frewshwater wetland ecosystems. In: Good RE, Whigham DF, Simpson RL, Simpson RL (eds) Freshwater wetlands: ecological processes and management potential. Academic Press, New York, pp 63–78

Gwin SE, Kentula ME, Shaffer PW (1999) Evaluating the effects of wetland regulation through hydrogeomorphic classification and landscape profiles. Wetlands 19:477–789

Hagen JB (1992) An entangled bank: the origins of ecosystem ecology. Rutgers University Press, New Brunswick, p 245

Hansson L, Brönmark AC, Nilsson PA, Åbjörnsson K (2005) Conflicting demands on wetland ecosystem services: nutrient retention, biodiversity or both? Freshw Biol 50:705–714

Kadlec RH, Bevis FB (2009) Wastewater treatment at the Houghton Lake wetland: vegetation response. Ecol Eng 35:1312–1332

Kandus P, Quintana RD, Minotti PG, Del Pilar Oddi J, Baigún C, González Trilla G, Ceballos D (2009) Ecosistemas de humedal y una perspectiva hidrogeomórfica como marco para la valoración ecológica de sus bienes y servicios. In: Laterra P, Jobbagy E, Paruelo J (eds) Valoración de Servicios Ecosistémicos: Conceptos, Herramientas y Aplicaciones para el Ordenamiento Territorial

Kantrud HA, Krapu GL, Swanson GA (1989) Prairie basin wetland of the Dakotas: a communtiy profile. United States Fish and Wildlife Service, Biological Report 85 (7.28), United States Government Printing Office, Washington, p 111

Keddy PA (2000) Wetland ecology: principles and conservation. Cambridge University Press, Cambridge, p 614

Kellison RC, Young MJ, Braham RR, Jones EJ (1998) Major alluvial floodplains. In: Messina MG, Conner WH (eds) Southern forested wetlands: ecology and management. Lewis Publishers, Boca Raton, pp 291–323

Kelly CA, Rudd JWM, Bodaly RA, Roulet NP, St. Louis VL, Heyes A, Moore TR, Schiff S, Aravena R, Scott KJ, Dyck B, Harris OR, Warner B, Edwards G (1997) Increases in fluxes of greenhouse gases and methyl mercury following flooding of an experimental reservoir. Environ Sci Technol 31:1334–1344

King SL (1995) Effects of flooding regimes on two impounded bottomland hardwood stands. Wetlands 15:272–284

Klimas CV, Murray EO, Langston H, Pagan J, Witsell T, Foti T (2008) A regional guidebook for conducting functional assessments of forested wetlands and riparian areas in the Ozark Mountains region of Arkansas. ERDC/EL TR-08-31, Environmental Laboratory, United States Army Engineer Research and Development Center, Vicksburg, p 200. http://el.erdc.usace.army.mil/wetlands/guidebooks.html

Kusler J (1992) Wetland delineation: an issue of science or politics. Environment 34:7–11, 29–37

Lugo AE, Snedaker SC (1974) The ecology of mangroves. Annu Rev Ecol Syst 5:39–64

Maltby E, Barker T, Linstead C (2009) Chapter 23. Development of a European methodology for the functional assessment of wetlands. In: Maltby E, Barker T (eds) The wetlands handbook. Wiley-Blackwell, Oxford, pp 486–512

Medina E, Cuevas E, Lugo AE (2009) Nutrient relations of dwarf *Rhizophora mangle* L. mangroves on peat in eastern Puerto Rico. Plant Ecol 207:13–24

Michener WK, Blood ER, Bildstein KL, Brinson MM, Gardner LR (1997) Climate change, hurricanes and tropical storms, and rising sea level in coastal wetlands. Ecol Appl 7:770–801

Millennium Ecosystem Assessment (2005) Ecosystems and human well-being: wetland and water synthesis. World Resources Institute, Washington, p 68

Mitsch WJ, Gosselink JG (2007) Wetlands, 4th edn. Wiley, New York, p 600

Muñoz-Reinoso JC (2001) Vegetation changes and groundwater abstraction in SW Doñana, Spain. J Hydrol 242:197–209

National Council for Air and Stream Improvement, Inc (NCASI) (1991) Forested wetlands classification and mapping: a literature review. Technical Bulletin No. 0606. National Council for Air and Stream Improvement, Inc., Research Triangle Park, North Carolina, p 99. http://www.ncasi.org//Publications/Detail.aspx?id=875

National Research Council (1995) Wetlands: characteristics and boundaries. National Academy Press, Washington, p 328

Nyman JA (2011) Ecological functions of wetlands. In: LePage BA (ed) Wetlands—integrating multidisciplinary concepts. Springer, Dordrecht

Odum HT, Pinkerton RC (1955) Time's speed regulator: the optimum efficiency for maximum power output in physical and biological systems. Am Sci 43:321–343

Odum HT, Copeland BJ, McMahan E (1974) Coastal ecosystems of the United States, vol 1–4. Conservation Foundation, Washington

Peterjohn WT, Correll DL (1984) Nutrient dynamics in an agricultural watershed: observations on the role of a riparian forest. Ecology 65:1466–1475

Richardson CJ (1983) Pocosins: vanishing wastelands or valuable wetlands? BioScience 33:626–633

Richter KO, Azous AL (1995) Amphibian occurrence and wetland characteristics in the Puget Sound Basin. Wetlands 15:305–312

Rodríguez JP, Beard TD Jr, Agard J, Bennett E, Cork S, Cumming G, Deane D, Dobson AP, Lodge DM, Mutale M, Nelson GC, Peterson GD, Ribeiro T (2005) Interactions among ecosystem services. In Carpenter SR, Pingali PL, Bennett EM, Zurek MB (eds) Ecosystems and human well-being: scenarios, vol 2. Findings of the scenarios working group, Millennium Ecosystem Assessment. Island Press, Washington, pp 431–448

Schlesinger WH (1977) Carbon balance in terrestrial detritus. Annu Rev Ecol Syst 8:51–81

Semeniuk CA, Semeniuk V (1995) A geomorphic approach to global wetland classification. Vegetatio 118:103–124

Semeniuk V, Semeniuk CA (1997) Geomorphic approach to global classification for natural inland wetlands and rationalization of the system used by the Ramsar Convention—a discussion. Wetlands Ecol Manage 5:145–158

Shaw SP, Fredine CG (1956) Wetlands of the United States. United States Department of the Interior, Fish and Wildlife Service Circular 39, Washington, p 67

Smith RD, Ammann A, Bartoldus C, Brinson MM (1995) An approach for assessing wetland functions using hydrogeomorphic classification, reference wetlands, and functional indices. United States Army Corps of Engineers, Waterways Experimental Station, Technical Report TR WRP-DE-10, Vicksburg, p 90. http://el.erdc.usace.army.mil/wetlands/wlpubs.html

Standler EK, Ehrenfeld JG (2009) Rapid assessment of urban wetlands: do hydrogeomorphoic classification and reference criteria work? Environ Manage 43:725–742

Warner BG, Rubec CDA (1997) The Canadian wetland classification system, 2nd edn. Wetlands Research Center, University of Waterloo, Ontario

Weller MW (1988) Issues and approaches in assessing cumulative impacts on waterbird habitat in wetlands. Environ Manage 12:695–701

Whittaker RH (1956) Vegetation of the great smoky mountains. Ecol Monogr 26:1–80
Williams CJ (2011) A paleoecological perspective on wetland restoration. In: LePage BA (ed) Wetlands—integrating multidisciplinary concepts. Springer, Dordrecht
Winter TC (1992) A physiographic and climatic framework for hydrologic studies of wetlands. In: Robarts RD, Bothwell ML (eds) Aquatic ecosystems in semi-arid regions: implications for resource management. Environment Canada NHRI Symposium Series 7, Saskatoon, pp 127–148

Chapter 6
Ecological Functions of Wetlands

John A. Nyman

Abstract We use the term "function" to mean processes or manifestations of processes. Wetlands probably have hundreds of functions, but only a small number are considered important to most people. Some values are based on aesthetics, such as beauty and naturalness, but many are economic, such as water quality improvement, wildlife hunting lease value, and sustainable timber production. Some of the values are considered to be negative, such as supporting mosquito populations. We classify functions as those that generally are captured by the landowner versus those that generally are external to the landowner to illuminate the fact that most of the economic value of most wetlands is external to the landowner.

We use the term "function" as the National Research Council (1995) broadly defined it as "processes or manifestations of processes." Wetlands probably have hundreds of functions (Table 6.1), but only a small number are considered to be important to most people because these are perceived to have value (i.e., they affect goods and services). In this chapter we focus on the functions that have values that fall into the "utility" system of wetland classification discussed by Brinson (2011). Some values are based on aesthetics, such as beauty and naturalness, but many are economic, such as water quality improvement, wildlife hunting lease value, and sustainable timber production. Some of the values are considered to be negative, such as supporting mosquito populations.

We classify functions differently than previous writers (Brinson 2011) in that the functions are classified as those that generally are captured by the landowner versus those that generally are external to the landowner. We classify wetland functions this way to illuminate the fact that most of the economic value of most wetlands is external to the landowner. Economically, this is an example of a market failure caused by an externality. Externality is a term that is used to describe

J. A. Nyman (✉)
School of Renewable Natural Resources, Louisiana State University Agricultural Center, Baton Rouge, LA 70803, USA
e-mail: jnyman@lsu.edu

B. A. LePage (ed.), *Wetlands,*
DOI 10.1007/978-94-007-0551-7_6, © Springer Science+Business Media B.V. 2011

Table 6.1 Wetland functions. (Modified from the National Research Council 1992)

Flood conveyance: Wetlands associated with rivers and floodplain often form natural flood-ways that convey floodwater from upstream to downstream sites

Protection from storm waves and erosion: Coastal wetlands and inland wetlands located adjacent to large lakes and rivers reduce the impact of storm tides and waves before they reach the uplands

Sediment control: Wetlands filter sediment from floodwaters; this function can become saturated

Habitat for fish and shellfish: Wetlands are essential spawning and nursery habitat that support commercial and personal fish and shellfish harvesting for human and animal food, especially in coastal waters, as well as for rare and endangered fish and shellfish

Habitat for waterbirds and other wildlife: Coastal and inland wetlands provide essential breeding, nesting, migratory, and non-breeding habitat for numerous species of waterfowl, wading birds, shorebirds, mammals, reptiles, and amphibians; some of these species are robust enough to support hunting whereas others are rare and endangered

Recreation: Wetlands are the sites for fishing, hunting, and non-consumptive use of wildlife and fish

Source of water supply: Wetlands are important in the transmission and quality of ground and surface water

Timber production: Under the proper management, forested wetlands are an important source of timber, but additional care will be needed in coastal wetland forests because of subsidence and/or sea-level rise

Preservation of historic and archaeological values: Some wetlands are of archeological interest, particularly those located near Native American settlements

Education and research: Wetlands provide significant educational opportunities for nature observation and scientific study

Open space and aesthetic value: Coastal and inland wetlands are areas of great diversity and beauty that provide open space for recreation and visual enjoyment

Water quality improvement: Wetlands contribute to improving water quality by removing excess nutrients and many chemical contaminants

a market failure that occurs generally when (1) the welfare of one person or firm is adversely affected by the action(s) of another person or firm, and (2) when the adversely affected party is not compensated (Gowdy and O'Hara 1995). However, this situation is reversed for many wetland landowners, who remain uncompensated by people and firms who benefit from the functions provided by the wetlands they own. The externalized benefit of wetland ownership is an obstacle to landowners who want to maintain wetlands and also motivates the conversion of wetlands into developed areas that can provide income. The general public benefits from privately owned wetlands, which is the rationale for government ownership of wetlands (e.g., wildlife management areas and water quality improvement areas) as well as the rationale for government regulation to restrict what private landowners can do with their wetlands (e.g., Section 404 of the Clean Water Act in the United States or the Water Framework Directive in the European Union). Some government programs partially correct this market failure by paying private landowners to restore or maintain their wetlands. Such programs unfortunately are often ridiculed as "paying farmers not to farm" when they really are systems in which society pays farmers for the benefits (ecological functions and societal values) provided by their wetlands, but captured primarily by others. In the United States, examples include the Conservation Reserve Program (CRP) and Wetland Reserve Program (WRP), which are operated by the United States Department of Agriculture (USDA).

Different types of wetlands provide a variety of functions. That is, not all wetlands provide the same types of ecological functions and societal values or if they do, the amounts are likely to be different. Quantifying how much function different wetlands provide is important for wetland managers and wetland regulators (Stein et al. 2010). Managers and regulators need such information when they evaluate the effectiveness of their management and restoration plans, when they consider which areas to purchase for inclusion in refuge systems or wetland mitigation banks, and when estimating the value of wetlands that will be destroyed by permitted activities. Until the late 1900's, wetland regulations in the United States failed to recognize the fact that some wetlands provide more functions than others. The concept of ranking wetlands by their ability to provide various functions now is widespread in the United States (Brinson and Rheinhardt 1996), Great Britain, and Europe (Maltby 2009). Unfortunately, quantifying such functions both ecologically and economically is easy for some, but fraught with uncertainty for others, especially for those functions whose value is external to the landowner.

Functions that Generally can Benefit the Landowner

Functions that benefit landowners generally are related to sustainable populations of plants, wildlife, and fish. The wildlife and fish species may be migrant, transitory, or resident. Understanding these functions require knowledge of what makes wetland plants different from the upland plants, and what separates a sustainable from an unsustainable harvest.

Wetland Plants

Wetlands and uplands generally contain structurally and compositionally different plant communities. Most upland plant species are generally excluded from wetlands because their roots cannot withstand prolonged periods of growth in soil that is flooded and devoid of air. Alternatively, most wetland plants do not require flooded soil to live, but they generally are excluded from the uplands by competition from upland plant species. The plant species found in wetlands can tolerate the stressful growing conditions created by flooded soil because of adaptations in their roots that deliver oxygen to the plant. This ability to supply oxygen to the plant root also is central to the water quality function of wetlands described later. Despite the stressful soil conditions associated with wetlands, wetlands do support highly productive plant communities that provide valuable functions.

When managed correctly, trees can be harvested from wetland forests on a sustainable basis. The practices required to sustainably harvest trees apparently were developed in Germany during the 1700s, spread to Europe, and by the late 1800s spread to the United States. Though such practices are now globally available, they are not universally practiced. Even without timber harvests, forest vegetation changes over time through cycles of disturbance and succession (the process of forest development that proceeds from a young or pioneer plant community to a mature or climax community). In mature forests, disturbance often results from the death of a single tree. In these cases sunlight that reaches the forest floor can initiate new cycles of succession that begin with grasses/groundcover which then proceeds through to shrubs, to juvenile trees, and finally to mature trees. Widespread disturbance occasionally occurs naturally in the form of fire, ice storms, hurricanes, and disease. Timber harvest prematurely initiates a new cycle of succession and often over large areas. Sustainable timber harvests can result in very unnatural systems that are analogous to tree farms with rows of trees that are harvested (i.e., disturbed in large patches every 40–60 years). Alternatively, sustainable timber harvests can result in ecosystems with many of the natural forest attributes if they are harvested in small patches less frequently. This form of forest management is more ecologically sound and appears to be more common in wetlands than the more intensive version.

Cutting and removing trees can decrease or increase the carrying capacity (the ability of a particular habitat to support a species population) for forest wildlife, with decreases more likely for wildlife species adapted to later stages of forest succession (i.e., mature forests). Traditional forest management was focused on the production of forest products such as lumber or pulp, but new forestry practices have been developed that retain enough commercially valuable timber to justify a commercial harvest while maintaining or creating the structural diversity that is valuable to numerous wildlife species, but especially to migrant songbirds. In many forests of the lower Mississippi Valley for example, the Lower Mississippi Valley Joint Venture Forest Resource Conservation Working Group (2007) recommended the following:

- An average of 60–70% overstory canopy cover that is heterogeneously distributed;
- A basal area of 60–70 ft² per acre, and 60–70% stocking;
- A desired 25–40% mid- and understory cover, respectively;
- At least two dominant tree species (emergents) per acre should be retained;
- Some cavity trees (small and large) as well as dead and/or stressed trees should be retained. These stems will eventually contribute to the coarse woody debris, which should average more than 200 ft³ per acre; and
- To ensure the future merchantability of the stands, shade-intolerant species regeneration should be present over 30–40% of the area.

Cutting and removing trees decreases water quality, primarily through increased soil erosion, but also through increased water temperature if the trees are harvested from near streams thereby allowing direct sunlight to reach the surface of the water. Best management practices (BMPs) have been recommended in some areas and required in others to minimize the effects of timber harvest on water quality (Loehle et al. 2009). Examples of BMPs include the creation and maintenance of a streamside management zone where the trees are not harvested and the construction of lateral or wing ditches that divert water from roadside ditches back into forests rather than down slope into streams (Louisiana Department of Agriculture and Forestry n.d.).

After the trees are harvested, natural regeneration from the existing seeds (including the seedbank) and stump sprouts often provide the source of trees for the newly developing forest, but in intensively managed forests or tree farms additional trees can be planted. Natural regeneration may fail in some coastal wetland forests where subsidence and sea-level rise have gradually created flooding conditions that are tolerable to mature trees, but prevent the germination and survival of seedlings (Faulkner et al. 2007). Under these conditions, a timber harvest that would otherwise be sustainable will be unsustainable and result in the conversion of the forested wetland into open water.

Wetland Wildlife and Fisheries

The most widely valued wetland function is providing habitat for wildlife and fish (National Research Council 1992). Some species of fish and wildlife depend upon wetlands for all or part of their life cycle, whereas other species merely are more abundant there than elsewhere. Examples of the former include species of waterfowl that globally depend on wetlands throughout their life cycle; an example of the latter are white-tailed deer in eastern North America, which occur in uplands, but reach greater density in the bottomland hardwood forests. Over one third of the United States federally endangered and threatened plants and wildlife require wetlands during some portion of their life cycle (National Research Council 1992).

In many parts of the world, landowners own all wildlife and fish on their property and so they can sell these resources. In North America however, individuals cannot own native wildlife and fish, which instead are the property of all citizens and held by the government in trust for the benefit of present and future citizens (Organ and Mahoney 2007). Landowners in North America nonetheless benefit economically from the wildlife and fish on their property because they can sell access to their property through hunting and fishing leases.

Many people mistakenly believe that pre-industrial or pre-agricultural humans sustainably harvested wildlife and fish, but there is little evidence of such situations except during the accidental combination of low human population numbers, inefficient technologies, and a lack of commercial markets (Hames 2007). On the other hand, examples abound demonstrating that the extinction of various wildlife species coincided with the arrival of humans as we spread across the planet (Steadman et al. 1995, 2005) and that as the human population increased, people switched their diets from a focus on slow moving, large species of wildlife that were easy to capture and that happened to reproduce slowly to a focus on species more difficult to capture, but who happened to reproduced faster (Stiner 2001). The practice required to sustainably harvest wildlife and fisheries apparently were known in medieval England

and Europe, but the benefits were limited to a few species and to only people born into nobility. Such practices did not become widespread until they were developed in the United States in the early 1900s and applied to all native species of wildlife and fish (Bolen and Robinson 2003). Currently, where wildlife and fisheries harvest is sustainable, it is because there are effective laws that limit the density of hunters and fishers in time as well as in space, limit the technologies used by hunters and fishers, and almost always prevent the commercial sale of harvested animals (Connelly et al. 2005). Unsustainable wildlife and fish harvest is problematic where the laws are not enforced or are based on biologically invalid concepts, or where technology and commercial markets fuel harvests that exceed rates of reproduction (Brashares et al. 2004).

Numerous species of wildlife and fish can be sustainably harvested from wetlands if the harvest rate does not exceed the reproduction rate. Examples include ungulates such as white-tailed deer and moose, rodents such as various species of squirrels, lagomorphs (rabbits), crustaceans such as crawfish, and birds such as waterfowl. In terms of the value to wetland owners, the most valuable are the waterfowl. Bellrose (1980) listed 55 species of ducks, geese, and swans in North America. Many waterfowl species breed during the summer near the poles, but within a few months they and their offspring will fly thousands of miles, many to the other hemisphere, to avoid the winter conditions developing in their breeding grounds (Bellrose 1980). Throughout their lives, most of which is spent traveling, they will depend on wetlands for food and shelter. The commercial hunting of waterfowl species ended in Canada, the United States, and Mexico during the early 1900s, but personal hunting continues for most waterfowl species throughout North, Central, and South America at certain times of the year. Europe has far fewer waterfowl species and those species are less abundant and consequently there is less hunting. Most waterfowl feed on seeds and invertebrates, but their dietary requirements change throughout the year, especially for females who must lay eggs. Hence most waterfowl management is focused on seed production by trees in forested wetlands and by grasses in marshes, which is generally accomplished by managing water levels to create the appropriate stresses at some times of the year and thereby eliminating the upland plants, while creating conditions that are favorable for the germination and growth of wetland plants at other times (Chabreck and Nyman 2005; Laubhan et al. 2005).

Carbon Storage

All wetlands store large amounts of carbon in the living plant tissues, but some wetlands store an even larger amount that is buried. The buried component is generally is known as peat. Peat wetlands occur throughout the world, but are most common throughout the boreal regions of the Northern Hemisphere and in coastal wetlands (Mitsch and Gosselink 2000). For centuries, this function had economic value only to the landowners who sold peat, often for fuel, but also as horticultural media.

However, using wetlands as sources of peat (i.e., peat mining) is an unsustainable activity and leaves the mined areas barren of plants and wildlife for decades. The burning of peat for home heating and commercial generation of electricity, and even its eventual decomposition in horticulture, ultimately releases the carbon (as carbon dioxide) into the atmosphere that had been stored for thousands of years. Not only does this contribute to global warming, but these activities change the former peat-land from an area that was actively storing carbon (carbon sink) to one that releases carbon (carbon source) (Cleary et al. 2005).

During the 1990's, a market developed for carbon credits worldwide. Demand in this new market allowed some landowners to earn money by selling the process by which wetlands accumulate peat (carbon sequestration) rather than by selling the end product (peat). Regarding wetlands, such efforts are almost exclusively associated with efforts to restore forested wetlands where they existed before they previously were cleared for agriculture rather than on marshes. Even in the United States, whose landowners were ineligible to sell carbon credits in the carbon trading system created by the Kyoto Protocol (because the United States failed to sign the Kyoto Protocol), landowners that could demonstrate carbon sequestration in their wetlands found buyers for their carbon credits because of numerous municipal, state, and regional policies that created alternative markets for carbon sequestration (Byrne et al. 2007).

Functions that Generally are External to the Landowner

Many of the functions that wetlands provide cannot be realized in the wetland and instead accrue to people living where the water, wildlife, and fish travel after leaving a particular wetland. Some of these functions are related to fish and wildlife, whereas others are related to water quality and flood control. Ingraham and Foster (2008) estimated that the total value of carbon sequestration, flood control, freshwater regulation and supply, water quality improvement, and habitat provision provided by wetlands in the National Wildlife Refuge System in the contiguous United States averaged $ 8,800 per acre per year ($ 3,561 per hectare per year) based on 2004 dollars. They also found that wetlands accounted for only 18% of area in the refuge system, but accounted for 85% of total value of the ecosystem services provided by the refuge system. We focus on what is thought to be the three most important wetland functions: improving water quality; supporting biodiversity including commercially important wildlife and fish; and floodwater storage.

Water Quality Improvement

Wetlands are so efficient at protecting and/or improving water quality that many municipalities find it less expensive to build a wetland than a chemical plant to remove excess nutrients from drinking water and sewage discharge. There is an entire industry devoted to the practice of constructing treatment wetlands (Kadlec and Knight 1996), and water quality accounted for the majority of the value reported by Ingraham and Foster (2008). The effect that wetlands have on nutrient availability in downstream water bodies is the most likely reason that wetlands in the United States are regulated by the federal government under the Clean Water Act. Understanding this function requires some understanding of phosphorus and nitrogen cycling. Both cycles include wetlands, but the cycles have one important difference; the nitrogen cycle has a gaseous phase, while the phosphorus cycle is limited to soil, water, and living organisms.

In most of the developed world, nutrients are major water pollutants (Smith 2003). The most important nutrients in aquatic ecosystems are nitrogen and phosphorus; thus contributions of even small amounts can have a significant impact on aquatic ecosystems. When nitrogen and phosphorus are at very low levels, as is common in most unpolluted water bodies, natural vegetation under the surface of the ponds, lakes, and coastal waters typically contain robust stands of rooted plants. These plants grow entirely underwater and are generally are referred to as Submersed Aquatic Vegetation (SAV). SAV provides valuable aquatic habitat, as food and shelter, to small fish and invertebrates. Water bodies that contain more SAV generally contain more fish and aquatic invertebrates (Kanouse et al. 2006) and hence more fish- and aquatic invertebrate eating birds, such as wading birds and waterfowl (O'Connell and Nyman 2010). Even minute increases in the availability of a limiting nutrient above natural levels can increase the growth of the SAV, and thus increase fish and wildlife abundance. However, it takes very little additional increase in the availability of phosphorous and/or nitrogen to alter the outcome of competition between SAV and phytoplankton.

Phytoplankton are single-celled plants that grow in virtually all water bodies. When nutrient availability is low as is the case in most natural water bodies, the number of phytoplankton (per unit of volume) is small and SAV dominates such ecosystems. As the nutrient availability increases, the phytoplankton gain a competitive advantage because phytoplankton depend entirely on the nutrients dissolved in the water. Given enough dissolved nutrients, the phytoplankton will come to dominate the ecosystem (Smith 2003). Under these conditions the phytoplankton can become so abundant that they color the water green and commonly eliminate the SAV by preventing the light from penetrating more than a few inches below the surface of the water. The phytoplankton effectively shade out the SAV. This condition is referred to as eutrophication. Eutrophic waterbodies can be more productive than unpolluted water bodies, but virtually all of the productivity is provided by the phytoplankton, which provides little ecological function or societal value. Fish and wildlife on the other hand, provide valuable functions and values in the form of non-consumptive recreation and sustainable hunting or fishing. Although such functions and values are present in eutrophic lakes, estuaries, and coastal waters, the relative value of these functions and values compared to pristine habitats is considerably less (Kemp et al. 2005). The decomposition of large amounts of microbial biomass on the bottom of lakes and even coastal waters can also deplete the dissolved oxygen, which then creates dead zones and eliminates commercially important fish species and their food sources (Rabalais et al. 2002).

Phosphorus is an important nutrient and water pollutant. Wetlands act as filters that intercept the phosphorus and remove it from the system before it can reach water bodies by adsorbing the phosphorous on clay particles (Mitsch and Gosselink 2000). Because of this function, wetlands can help to prevent or minimize eutrophication and are commonly used to reverse eutrophication in polluted lakes (Coveney et al. 2002). However, wetlands have a finite capacity to hold soil particles, and soil particles have a finite capacity to hold phosphorus. When artificial wetlands have reached their capacity to hold the phosphorus, they are mined and the soil can be used as fertilizer.

Nitrogen is an important nutrient and water pollutant and wetlands also act as filters for nitrogen. But more importantly they have an almost unlimited capacity to convert the nitrogen from forms that are available to plants to nitrogen gas that is unavailable to plants.

During the 1800s, scientists had learned about parts of the nitrogen cycle that returned what was then known as combined nitrogen. Today we say fixed nitrogen because the nitrogen is chemically bound or "fixed," to a different element such as in nitrate (NO_3^-) or ammonia (NH_4^+) to the atmosphere. In addition, the early scientists had not yet learned that there are symbiotic bacteria living in the roots of some plants that converted nitrogen gas into forms of nitrogen that are available for plant uptake (Kellerman 1911). Thus, during the 1800s some scientists calculated the probable date of exhaustion of the world's supply of combined nitrogen and the subsequent starvation of all living beings on earth (Kellerman 1911). In the 1890s and early 1900s however, scientists learned of the natural processes of nitrogen fixation. The primary process responsible for the loss of fixed nitrogen from the landscape is denitrification (Fig. 6.1). Denitrification contributes greatly to wetlands removing nitrogen more efficiently than other habitat types (Seitzinger 1994, 2006).

Denitrification is the microbially mediated conversion of nitrate-N (NO_3–N) to elemental nitrogen (N_2) and nitrous oxide (N_2O). Nitrate is soluble in water and is readily available to plants whereas N_2 and N_2O are gases and unavailable to plants (Seitzinger et al. 2006). The necessary conditions for denitrification are found at the interface in the soil where oxic (with oxygen) and anoxic conditions (without oxygen) exist. In most ecosystems, these conditions are found only at the ground or sediment surface. The reason that wetlands are hotspots for denitrification is that the wetland plants have roots that oxidize the soil adjacent to their roots thereby creating an immense oxic and anoxic interface throughout the soil that greatly exceeds the interface area between oxic and anoxic conditions at the soil surface. It is irrelevant whether the water flow is above or belowground; thus even wetlands that lack a surface connection to rivers and stream may be removing large amounts of nitrogen from groundwater before the groundwater reaches streams, lakes, and coastal waters (Whigham and Jordan 2003).

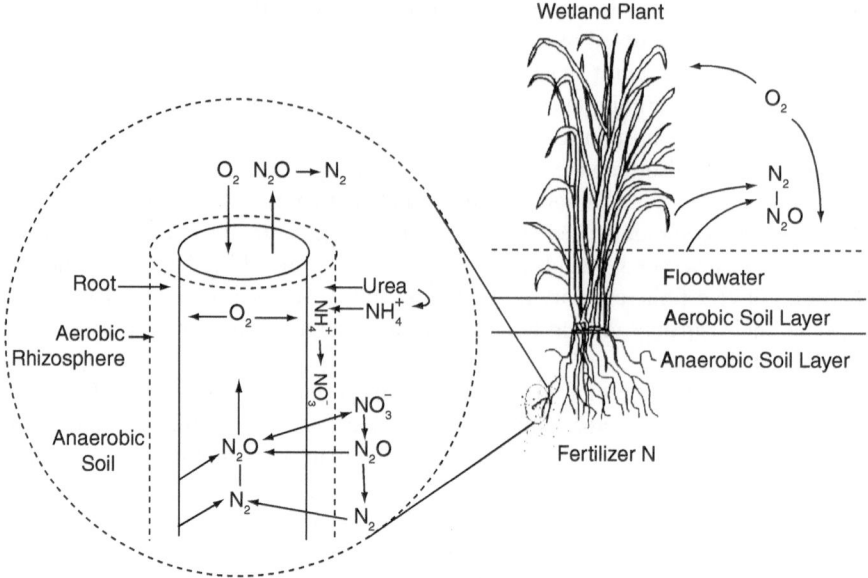

Fig. 6.1 Schematic presentation of nitrification-denitrification in the root zone of vascular wetland plant species. (From Reddy et al. 1989)

Wildlife and Fisheries

As mentioned in the preceding section, the most widely valued wetland function is providing habitat for wildlife and fish (National Research Council 1992). Migratory birds depend on or use wetlands for part of their life cycle, but the value of this function extends across countries and continents. As noted earlier, many species of waterfowl fit this pattern and thus the economic value of a wetland often is shared across countries and continents. Such economic value of wetlands is reflected in the ecotourism dollars that are generated in towns located near wildlife refuges (Caudill and Henderson 2005). Most of these species however are not (legally) hunted, but they are valuable for other, primarily non-economic reasons. Hundreds of species of songbirds for example depend on forested wetlands that are spread across thousands of miles. Likewise, hundreds of species of shorebirds depend upon coastal wetlands and interior marshes that are located thousands of miles apart. The seasonal function of wetlands is the reason that so many wildlife refuges contain wetlands. Except for the brief period each year during migration, these wetlands appear to be unimportant to humans; they are nevertheless of crucial importance to the thousands of animals that use them.

Like wildlife, there are many fish that use wetlands only during part of their life cycle, but even so, these species are wetland dependent. Perhaps the most economically valuable fish species in the United States is the menhaden, which is caught along the Atlantic and Gulf Coasts (Franklin 2007). People consume this fish only indirectly because it is used primarily as feed for poultry and pigs, but pets consume it directly. Menhaden depend on coastal wetlands to complete their first year of life (Deegan et al. 1990). Years in which coastal wetlands are more productive because of annual variations in sea level also produce more menhaden (Morris et al. 1990). As coastal wetlands are developed or converted to shallow open water because of subsidence (a drop in the land surface) and/or sea-level rise (Kearney and Stevenson 1991; Boesch et al. 1994), the carrying capacity of coastal wetlands to support menhaden populations will decline and menhaden populations will decrease, assuming that they are not first overfished. As menhaden become less abundant, the consumers of the makers of animal feed will be forced find alternatives, such as soy. However, if those alternative sources were less expensive, they would be in use already. The owners of the wetlands currently supporting menhaden populations are not compensated for this function by meat eaters and pet owners who benefit from it, but it is certain that coastal wetland loss will lead to slight increases in the prices paid by pet owners and meat eaters as menhaden become too rare or too expensive to commercially harvest. Menhaden are only one example of wetland dependent fish.

Floodwater Storage

Storing floodwater is an exclusively economic value. Wetlands have long been recognized for their value to store floodwaters and thereby reduce the height of floods and hence the economic damage caused by river or oceanic flooding. This function is another that is not expressed in the wetland itself. Instead, the effect of this function accrues downstream of the floodwaters. This is true for wetlands associated with rivers, as well as coastal wetlands. In the case of coastal wetlands, downstream refers to the movement of storm surge from the ocean onto the land.

Wetland Functions and Landscape Patterns

The location of a wetland relative to other habitat types on the landscape influences its ability to provide various functions. For example, wetlands located adjacent to streams in agricultural landscapes generally remove more nitrogen than groundwater wetlands in forested landscapes. Likewise, the loss of small isolated wetlands in some landscapes would reduce the wetland area only 19%, but would reduce the area of habitat suitable for terrestrial-dwelling, aquatic-breeding amphibians from 90% to 54% of the landscape (Gibbs 1993).

The landscape pattern within wetlands also affect the level at which wetlands provide various ecological functions. At this scale, the interspersion of open water and emergent vegetation within wetlands greatly affects the capacity at which wetlands can improve water quality (Thullen et al. 2005), provide breeding habitat for waterfowl (Fairbairn and Dinsmore 2001), provide wintering habitat for wading birds and waterfowl (O'Connell and Nyman 2010), and provide breeding habitat for mosquitoes (Thullen et al. 2005). Interestingly, for all these functions, the optimal level of function (i.e., all high except mosquito production), occurs when there is high amount of edge between emergent vegetation and open water (Fairbairn and Dinsmore 2001; Thullen et al. 2005; O'Connell and Nyman 2010).

Conclusions

Wetlands can certainly improve water quality, but because the phosphorus cycle lacks a gaseous phase and the nitrogen has a gaseous phase, wetlands have a finite capacity to protect receiving waters from possessing high levels of phosphorus, but they have an almost infinite capacity to protect the receiving waters from excess nitrogen. Likewise, wetlands sustain populations of hundreds of species of wildlife and fish, many of which are migratory and thus depend on wetlands that are often spread across countries and continents. Despite these tremendously valuable functions, they remain external to most wetlands at most times. Few wetland owners are able to capture the economic benefits provided by their wetlands because those benefits have traditionally been limited to sustainable timber harvest and hunting leases. A growing market for carbon credits may add economic value to an ecological function that has been traditionally economically unimportant to wetland landowners and managers.

References

Bellrose FC (1980) Ducks, geese, and swans of North America, 3rd edn. Wildlife Management Institute. Stakepole Books, Harrrisburg, p 540

Boesch DF, Josselyn MN, Mehta AJ, Morris JT, Nuttle WK, Simenstad CA, Swift D (1994) Scientific assessment of coastal wetland loss, restoration and management in Louisiana. J Coastal Res (Special Issue) 20:1–103

Bolen EG, Robinson WL (2003) Wildlife ecology and management, 5th edn. Benjamin Cummings, New Jersey, p 634

Brashares JS, Arcese P, Sam MK, Coppolillo PB, Sinclair ARE, Balmford A (2004) Bushmeat hunting, wildlife declines, and fish supply in West Africa. Science 306:1180–1183

Brinson MM (2011) Classification of wetlands. In: LePage BA (ed) Wetlands—integrating multidisciplinary concepts. Springer, Dordrecht

Brinson MM, Rheinhardt R (1996) The role of reference wetlands in functional assessment and mitigation. Ecol Appl 6:69–76

Byrne J, Hughes K, Rickerson W, Kurgelashvili L (2007) American policy conflict in the greenhouse: divergent trends in federal, regional, state, and local green energy and climate change policy. Energy Policy 35:4555–4573

Caudill J, Henderson E (2005) Banking on nature 2004: the economic benefit to local communities of National Wildlife Refuge visitation. Division of Economics, United States Fish and Wildlife Service, Washington DC, p 435

Chabreck RH, Nyman JA (2005) Managing coastal wetlands for wildlife. In: Braun CE (ed) Techniques for wildlife investigations and management. The Wildlife Society, Bethesda, pp 839–860

Cleary J, Roulet NT, Moore TR (2005) Greenhouse gas emissions from Canadian peat extraction, 1990–2000: a life-cycle analysis. Ambio 34:456–461

Connelly JW, Gammonley JH, Peek JM (2005) Harvest management. In: Braun CE (ed) Techniques for wildlife investigations and management. The Wildlife Society, Bethesda, pp 658–690

Coveney MF, Stites DL, Lowe EF, Battoe LE, Conrow R (2002) Nutrient removal from eutrophic lake water by wetland filtration. Ecol Eng 19:141–159

Deegan LA, Peterson BJ, Potier R (1990) Stable isotopes and cellulase activity as evidence for detritus as a food source for juvenile Gulf menhaden. Estuaries 13:14–19

Fairbairn SE, Dinsmore JJ (2001) Local and landscape-level influences on wetland bird communities of the prairie pothole region of Iowa, USA. Wetlands 21:41–47

Faulkner SP, Chambers JL, Conner WH, Keim RF, Day JW, Gardiner ES, Hughes MS, King SL, McLeod KW, Miller CA, Nyman JA, Shaffer GP (2007) Conservation and use of coastal wetland forests in Louisiana. In: Conner WH, Doyle TW, Krauss KW (eds) The ecology of tidal freshwater swamps of the southeastern United States. Springer, The Netherlands, pp 447–460

Franklin HB (2007) The most important fish in the sea: menhaden and America. Island Press, Washington, p 265

Gibbs JP (1993) Importance of small wetland for the persistence of local populations of wetland-associated animals. Wetlands 13:25–31

Gowdy J, O'Hara S (1995) Economic theory for environmentalists. St. Lucie Press, Delray Beach, p 192

Hames R (2007) The ecologically noble savage debate. Annu Rev Anthropol 36:177–190

Ingraham MW, Foster SG (2008) The value of ecosystem services provided by the U.S. National Refuge System in the contiguous U.S. Ecol Econ 67:608–618

Kadlec RH, Knight RK (1996) Treatment wetlands. CRC Press, Boca Raton, p 893

Kanouse S, La Peyre MK, Nyman JA (2006) Nekton use of *Ruppia maritima* and non-vegetated bottom habitat types within brackish marsh ponds. Mar Ecol Progr Ser 327:61–69

Kearney MS, Stevenson JC (1991) Island land loss and marsh vertical accretion rate evidence for historical sea level changes in Chesapeake Bay. J Coast Res 7:403–415

Kellerman KF (1911) Nitrogen gathering plants. In: Arnold JA (ed) Yearbook of the United States Department of Agriculture 1910. Kessinger Publishing, Washington, pp 213–218

Kemp WM, Boynton WR, Adolf JE, Boesch DF, Boicourt WC, Brush G, Cornwell JC, Fisher TR, Glibert PM, Hagy JD, Harding LW, Houde ED, Kimmel DG, Miller WD, Newell RIE, Roman MR, Smith EM, Stevenson JC (2005) Eutrophication of Chesapeake Bay: historical trends and ecological interactions. Mar Ecol Progr Ser 303:1–29

Laubhan MK, King SL, Fredrickson LH (2005) Managing inland wetlands for wildlife. In: Braun CE (ed) Techniques for wildlife investigations and management, 6th edn. The Wildlife Society, Bethesda, pp 797–838

Loehle C, Wigley TB, Schilling E, Tatum V, Beebe J, Vance E, van Deusen P, Weatherford P (2009) Achieving conservation goals in managed forests of the southeastern coastal plain. Environ Manag 44:1136–1148

Louisiana Department of Agriculture and Forestry (n.d.) Recommended forestry best management practices for Louisiana. Louisiana Department of Agriculture and Forestry, Baton Rouge, Louisiana, p 84

Lower Mississippi Valley Joint Venture Forest Resource Conservation Working Group (2007) Restoration, management, and monitoring of forest resources in the Mississippi alluvial valley: recommendations for enhancing wildlife habitat. In: Wilson R, Ribbeck K, King S, Twedt D (eds). Lower Mississippi Valley Joint Venture Office, Vicksburg, p 96

Maltby E (2009) Functional assessment of wetlands: towards evaluation of ecosystem services. CRC Press, Boca Raton, p 712

Mitsch WJ, Gosselink JG (2000) Wetlands, 3rd edn. Wiley, New York, p 920

Morris JT, Kjerfve B, Dean JM (1990) Dependence of estuarine productivity on anomalies in mean sea level. Limnol Oceanogr 35:926–930

National Research Council (1992) Restoration of aquatic ecosystems: science, technology, and public policy. National Research Council, National Academy of Sciences. National Academy Press, Washington, p 576

National Research Council (1995) Wetlands: characteristics and boundaries. National Research Council, National Academy of Sciences. National Academy Press, Washington, p 328

O'Connell JL, Nyman JA (2010) Marsh terraces in coastal Louisiana increase marsh edge and densities of waterbirds. Wetlands 30:125–135

Organ J, Mahoney S (2007) The future of public trust. Wildl Prof 1:18–22

Rabalais NN, Turner RE, Wiseman WJ Jr (2002) Gulf of Mexico hypoxia, a.k.a. "the dead zone". Annu Rev Ecol Syst 33:235–263

Reddy KR, Patrick WH Jr, Lindau CW (1989) Nitrification-denitrification at the plant root-sediment interface in wetlands. Limnol Oceanogr 34:1004–1013

Seitzinger S (1994) Linkages between organic matter mineralization and denitrification in eight riparian wetlands. Biogeochemistry 25:19–39

Seitzinger SP, Harrison JA, Bohlke JK, Bouwman AF, Lowrance R, Peterson B, Tobias C, Van Drecht G (2006) Denitrification across landscapes and waterscapes: a synthesis. Ecol Appl 16:2064–2090

Smith V (2003) Eutrophication of freshwater and coastal marine ecosystems a global phenomena. Environ Sci Pollut Res 10:126–139

Steadman DW (1995) Prehistoric extinctions of pacific island birds—biodiversity meets zooarchaeology. Science 267:1123–1131

Steadman DW, Martin PS, MacPhee RDE, Jull AJT, McDonald HG, Woods CA, Iturralde-Vinent M, Hodgins GWL (2005) Asynchronous extinction of later quaternary sloths on continents and islands. Proc Natl Acad Sci 102:11763–11768

Stein ED, Brinson M, Rains MC, Lleindl W, Hauer FR (2010) Wetland assessment alphabet soup: how to choose (or not choose) the right assessment method. Wetl Sci Pract 26:20–24

Stiner MC (2001) Thirty years on the "Broad Spectrum Revolution" and paleolithic demography. Proc Natl Acad Sci 98:6993–6996

Thullen JS, Sartoris JJ, Nelson SM (2005) Managing vegetation in surface-flow wastewater-treatment wetlands for optimal treatment performance. Ecol Eng 25:583–593

Whigham DF, Jordan TE (2003) Isolated wetlands and water quality. Wetlands 15:541–549

Chapter 7
Multidisciplinary Approaches to Climate Change Questions

Beth A. Middleton

Abstract Multidisciplinary approaches are required to address the complex environmental problems of our time. Solutions to climate change problems are good examples of situations requiring complex syntheses of ideas from a vast set of disciplines including science, engineering, social science, and the humanities. Unfortunately, most ecologists have narrow training, and are not equipped to bring their environmental skills to the table with interdisciplinary teams to help solve multidisciplinary problems. To address this problem, new graduate training programs and workshops sponsored by various organizations are providing opportunities for scientists and others to learn to work together in multidisciplinary teams. Two examples of training in multidisciplinary thinking include those organized by the Santa Fe Institute and Dahlem Workshops. In addition, many interdisciplinary programs have had successes in providing insight into climate change problems including the International Panel on Climate Change, the Joint North American Carbon Program, the National Academy of Science Research Grand Challenges Initiatives, and the National Academy of Science. These programs and initiatives have had some notable success in outlining some of the problems and solutions to climate change. Scientists who can offer their specialized expertise to interdisciplinary teams will be more successful in helping to solve the complex problems related to climate change.

Multidisciplinary approaches are crucial to more fully understanding the interwoven environmental and social patterns of past and future episodes of climate change. An interdisciplinary analysis of these complex patterns is particularly valuable to inform models of the effects of future climate change on environments and humans. Historically, large climate shifts often have accompanied the rise and fall of civiliza-

---

Any use of trade, product, or firm names is for descriptive purposes only and does not imply endorsement by the United States Government.

B. A. Middleton (✉)
National Wetlands Research Center, United States Geological Survey, Lafayette, LA 70506, USA
e-mail: middletonb@usgs.gov

B. A. LePage (ed.), *Wetlands,*
DOI 10.1007/978-94-007-0551-7_7, © Springer Science+Business Media B.V. 2011

tions (Constanza et al. 2007), and in such circumstances, sediment records indicate that civilizations such as the Bal He Kuk Kingdom (Korea), Old Kingdom Egypt (Egypt), and the Maya (Yucatan Peninsula, Mexico) can fall within a few years (Redman et al. 2007). Thus, historical and paleoecological records of past human-climate interactions comprise important data for mathematical simulation models of future scenarios of the effects of climate change and their effects on human interactions with the environment (Dearing 2007). The validity of future climate scenarios based on simulation modeling requires an interdisciplinary melding of ideas from scientists spanning a wide range of disciplines from history to physics (Constanza et al. 2007). Despite this clear need for an interdisciplinary approach to climate change questions, few scientists are equipped to effectively deal with multidisciplinary questions.

Successful solutions to climate change problems must be formulated across many scientific disciplines, but the current training of ecologists is far too narrow for most of them to effectively enter an interdisciplinary dialogue. From just the perspective of ecology, climate change will have interwoven effects on ecosystems, biodiversity, hydrologic systems, and pathogens at many different spatial and temporal scales (National Academy of Science 2001). Such overarching ecological questions require scientists with broad training including biological engineers, physical ecologists, and historical biologists to adequately address interdisciplinary questions. Despite this need, individuals with broad training and/or a willingness to embrace multidisciplinary research questions are rare. We need scientists who are able to work together to see the consequences of potential global change solutions outside of the boundaries of narrow disciplines. For example, the unintended consequences of projects to remove massive amounts of CO_2 from the atmosphere by geoengineers, or to assist the migration of slow moving species by biologists. At the same time, funding agencies tend to favor projects posing reductionist questions. Needless to say, funding is the biggest driving force behind the researchers and research questions being undertaken at the local, national, and international level. Scientists would be more compelled to look at multidisciplinary questions using new research paradigms, if funding were widely available to support these teams.

Multidisciplinary Training Opportunities

Despite the clear need for an interdisciplinary approach to climate change questions, most ecologists lack the mindset and training to work efficiently as part of an interdisciplinary research team (see LePage 2011). Reductionists lack training via interdisciplinary degrees, and even with a willingness to work on multidisciplinary teams, narrowly trained scientists may not immediately comprehend either the broad perspectives of the interdisciplinary research question or the narrower focus of other interdisciplinary team members. This can make the coordination efforts of interdisciplinary research problems difficult, so that orientation meetings of interdisciplinary collaborative projects are useful.

Fortunately, some new graduate and workshop opportunities attempt to train scientists to become better interdisciplinary thinkers and collaborators. One example of an academic cross-disciplinary program is the Cluster Hiring Initiative of the University of Wisconsin (2008). This program was started in 1998 to facilitate interdisciplinary training, research, and collaborative writing projects, although it is not specifically geared toward climate change questions. The Cluster Hiring Initiative supports the hiring of 144 interdisciplinary faculty in 49 clusters, with some of the environmental clusters spanning a wide intellectual swath within subdisciplines related to agroecology, chemical biology, international environmental affairs and global security, genomics, landuse, molecular biometry, science and technology studies, and structural biology. Universities that provide interdisciplinary training in the natural, physical, and social sciences are very few in the United States (National Academy of Science 2001), but a wider availability of such programs may increase the numbers of scientists equipped to tackle interdisciplinary questions of relevance to climate change questions.

Workshops and conferences also provide a training avenue in multidisciplinary research perspectives. For example, the Santa Fe Institute is a research and training center for the education of students, faculty, and professionals (Santa Fe Institute 2008). Their workshops support multidisciplinary research and teaching collaborations that bridge physical, biological, computational, and social sciences, to better understanding the linkages of complex systems. Current research topics support the development of interdisciplinary thinking skills including: (1) physics of complex systems; (2) emergence, innovation, and robustness in evolutionary systems; (3) information processing and computation in complex systems; (4) dynamics and quantitative studies of human behavior; and, (5) emergence, organization, and dynamics of living systems (Santa Fe Institute 2008). The Dahlem Workshops are a European counterpart, and these conferences are organized to promote discussion, collaborative research, and effective communication between scientists engaged in interdisciplinary research (Constanza et al. 2007). Dahlem Workshops have been organized by the German Research Foundation since 1974 (Dahlem Konferenzen 2007).

Interdisciplinary Projects in Global Change Ecology

International Panel on Climate Change (IPCC)

Despite the problems with training and funding of multidisciplinary research projects, a number of very successful global change projects of an interdisciplinary nature can be reported. Perhaps the most famous global change project is that of the International Panel on Climate Change (IPCC), whose efforts represent a major international initiative to synthesize climate impact information to project the effects of climate change on natural and human systems (Bates et al. 2008). In general, the IPCC models drive the discussion of various teams of IPCC scientists to make projections of the impact of climate change on natural and human systems. One IPCC writing team focuses on freshwater resources, and they have considered climate projections based on models of climate change (Bates et al. 2008). The objectives of the freshwater resources writing group include: (1) understanding the links between natural and anthropogenic climate change, and options for mitigating water related problems; and, (2) conveying their findings to policymakers and stakeholders. The work of the IPCC stands as a testimony to the importance of interdisciplinary thinking and collaboration.

Joint North American Carbon Program

The Joint North American Carbon Program (JNACP) is a coalition of scientists working to understand the complex interrelationships between the carbon cycle and climate system of North America and adjacent coastal seas. The interdisciplinary scientists involved in this project include atmospheric scientists, economics, sociologists, and field ecologists. These researchers work in integrated teams to study spatial and temporal distribution of carbon sources, sinks, and greenhouse gas emissions. Their research questions examine the effects of climate change, natural disturbance, and the socioeconomic and institutional drivers of the carbon cycle (pools, fluxes, and processes), as well as climate scenarios based on concentrations of greenhouse gases (CO_2, CO, CH_4, and N_2O). Social scientists and economists explore climate change outcomes as projected from potential policies, management strategies, and new technologies designed to shift emissions of greenhouse gases or increase carbon pools.

National Academy of Science Research Initiatives

The National Academy of Sciences has outlined a number of "Grand Challenges", which are multidisciplinary questions of interest to scientists working on climate change (National Academy of Science 2003). These "Grand Challenges" include studies to increase the understanding of climate change in relation to: (1) biogeochemical cycles and how these are influenced by human activities; (2) biological diversity and ecosystem functioning at all spatial and temporal scales; (3) climate variability; (4) hydrological forecasting; (5) infectious diseases and their relationship to environment; (6) human resource usage as shaped by human institutions such as markets, governments, and treaties; (7) landuse dynamics, ecosystem function, and human welfare; and, (8) material usage and its potential recycling. All aspects of the environment are interconnected, so that a directive by the NAS to study such big-minded approaches unifies efforts toward that purpose.

Recognizing the limited opportunities of scientists and managers to communicate the findings of their interdisciplinary research, the NAS recommended that the National Science Foundation (NSF) and other agencies sponsor workshops for scientists, managers, and users of information to discuss research and implementation of the research findings (National Academy of Science 2003). These initiatives speak toward the fact that recognition is growing within the major research engines, that interdisciplinary research is necessary to answer our most pressing environmental questions, and that this information needs to be available to managers and the public.

The United States Geological Survey (USGS) has developed climate change initiatives following these NAS guidelines. The USGS addresses geological, hydrological, geographical, and biological process questions, and disseminates that information to the public. Long term monitoring is one of the USGS's most important contributions to science (National Academy of Science 2001), and in fact, most universities and other entities are not as well positioned to support this type of long-term research. The USGS (Frondorf et al. 2005) has allied its future science directions with the NAS's (2003) Grand Challenges including the: (1) study of the complexity of environmental phenomena; (2) capacity building for interdisciplinary research; and (3) the need to improve the relevance of environmental science research to problems of interest to the public (Frondorf et al. 2005). After a study requested by USGS, the NAS (2001) recommended that the USGS put an emphasis on multi-scale, multidisciplinary research toward addressing societal needs to solve questions related to highly complex environmental problems, following their own guidelines and vision of overall research direction.

Some Multidisciplinary Research Questions for Future Climate Change Research

Assisted Migration

Any number of multidisciplinary research questions could be of value to address the difficult challenges we may face in the future as related to climate change. Among many pressing ecological issues, with little supporting research, regards the ability of species to migrate and establish beyond their current distributional ranges should the climate become too hot or dry for these species southward. Assisted migration studies would inform managers as to the efficacy of physically moving species northward to novel locales, and testing the ability of these species to grow and regenerate (sexually and vegetatively) under the current climate regime. To accomplish these studies related to biological conservation, teams of interdisciplinary researchers including restorationists, managers, science historians, modelers, and ethicists need to be assembled to focus their attention on how species may be conserved under various climate change scenarios. In particular, natural philosophers, ecologists, and historians need to consider the ethics of moving species outside of their pre-settlement range. The idea of moving alien species into northern regions may be met by managers and the general public with resistance, and the inevitability and advisability of these procedures need to be evaluated. Also to support any efforts toward assisted migration, ecologists and climatologists will be required to study the mechanics of future climates on existing species and ecotypes. This empirical information is important for models of species responses to novel environments. All of these people will be required to "think outside of the box" because these questions cannot be addressed using anything but pieces of existing research and management paradigms.

Long-term Seed Storage

Because the world's seeds are at risk from climate change and other threats, a variety of long-term seed storage strategies including seed storage vaults and centers for genetic preservation have been designed through the efforts of problem solvers with insight into agricultural, ecological, and engineering solutions (Global Crop Diversity Trust 2009; United States Department of Agriculture 2009). The purpose of these centers is to preserve the genetic diversity of crop and sometimes wild species. Seeds can be stored in permafrost and thick rock indefinitely in the Svalbard Arctic Seed Vault. Svalbard is a group of islands about 1,000 km north of Norway (Global Crop Diversity Trust 2009) (Fig. 7.1).

Geoengineering Solutions

Among the climate change solutions under discussion include geoengineering of the planet (e.g., physically cooling the Earth or removing CO_2 from the atmosphere). While especially engineers may welcome these ideas as the solution to

Fig. 7.1 Svalbard global seed vault, Svalbard, Norway. (Photo by Mari Tefre: www. regjeringen.no/en/dep/lmd/ campain/svalbard-global-seed-vault/picture-archive. html?id=462226)

climate change problems, others including ecologists may warn of unanticipated consequences of these global scale projects. Any engineering solution requires thorough interdisciplinary discussion among engineers, climate scientists, ecologists, and ethicists, particularly because of the potentially far-reaching effects of these projects on human society and the environment.

A number of examples of potential geoengineering projects are described on the website by ChooseClimate.org (2007). Ideas to cool the Earth directly include: (1) engineering weather to create rain and/or cloudy weather to cool the Earth via cloud seeding or storm diversion; (2) the addition of sulphate aerosols or dust in the atmosphere; (3) giant reflectors placed into orbit around the Earth to reflect sunlight; (4) changing various landscapes to have a lighter colored albedo to reflect light from the Earth (e.g., changing color of the desert to white), and; (5) the subduction of cold water in the ocean. A number of geoengineering ideas to cool the planet seek to remove CO_2 from the atmosphere including the stimulation of ocean algae in the ocean by fertilization with nitrate and phosphate, or adding iron to ocean water to support the growth of kelp farms, because iron may be a key limiting micronutrient for kelp. Tree plantations might also remove CO_2 from the atmosphere, because young trees take up CO_2, but not necessarily older trees. Watering and/or fertilizing the desert have been proposed to increase the CO_2 uptake of desert plant species, but greening the desert would change the albedo of the surface of the desert, and could cause these landscapes to heat up. Other proposals include adding limestone to the ocean to fix CO_2, or to storing CO_2 under rocks (ChooseClimate.Org 2007). Perhaps the geoengineering solution that has received the most attention has been for specific research designed to remove CO_2 from the atmosphere and return it to the ocean for storage (House et al. 2007). These geoengineering projects typically do not include scientists who are not engineers, and will obviously have many other direct and indirect consequences for the environment in addition to cooling the Earth or removing atmospheric CO_2. For all such encompassing projects to be successful, a thorough interdisciplinary discussion is required.

Concluding Statements

Interdisciplinary approaches are necessary to solve the novel problems related to climate change. The NAS recognizes the need for big-minded approaches for the study of climate change, and has called scientists to undertake interdisciplinary research to answer these questions. No one scientist can have the expertise to tackle extremely complex questions, and we must learn to better assimilate concepts and work as a team to solve complex problems. These interdisciplinary skills are especially important for ecologists to develop strategies to approach novel problems related to climate change impacts on interlinked environmental and human systems.

References

Bates B, Kundzewicz ZW, Wu S, Palutikof J (2008) Climate change and water. Intergovernmental Panel on Climate Change. Technical paper of the Intergovernmental Panel on Climate Change. IPCC Secretariat, Geneva, Switzerland, p 210

ChooseClimate.Org (2007) Part 2. Climate engineering proposals. http://www.chooseclimate.org/cleng/part2.html. Accessed 9 Jan 2009

Constanza R, Graumlich LJ, Steffen W (2007) Sustainability or collapse: lesson from integrating the history of humans and the rest of nature. In: Constanza R, Graumlich LJ, Steffen W (eds) Sustainability or collapse: an integrated history and future of people on earth. MIT Press, Cambridge, pp 3–17

Dahlem Konferenzen (2007) History of the Dahlem Konferenzen. Frei Universität, Berlin, Germany. http://www.fu-berlin.de/veranstaltungen/dahlemkonferenzen/en/geschichte/index.html. Accessed 12 Jan 2009

Dearing, JA (2007) Human-environment interactions: learning from the past. In: Constanza R, Graumlich LJ, Steffen W (eds) Sustainability or collapse: an integrated history and future of people on earth. MIT Press, Cambridge, pp 19–48

Frondorf AF, Boldt DR, Hutchison DR, Miller WG, Posson DR, Sipkin SA (2005) Future science direction: environmental information science plan. Open-file Report 2005-1180. United States Department of the Interior, United States Geological Survey, Reston, VA

Global Crop Diversity Trust (2009) Arctic seed vault. Svalbard Global Seed Vault, Global Crop Diversity Trust, Svalbard

House KZ, House CH, Schrag DP, Aziz MJ (2007) Electrochemical acceleration of chemical weathering as an energetically feasible approach to mitigating anthropogenic climate change. Environ Sci Technol 41:8464–8470

LePage BA (2011) Wetlands: a multidisciplinary perspective. In: LePage BA (ed) Wetlands—integrating multidisciplinary concepts. Springer, Dordrecht

National Academy of Science (2001) Future roles and opportunities for the United States Geological Survey. National Academy Press, Washington, p 192

National Academy of Science (2003) Grand challenges in environmental sciences. National Research Press, Washington, p 106

Redman DL, Crumley CL, Hassan FA, Hole F, Morais J, Riedel F, Scarborough VL, Tainter JA, Turchin P, Yasuda Y (2007) Group report: millennial perspectives on the dynamic interaction of climate, people, and resources. In: Constanza R, Graumlich LJ, Steffen W (eds) Sustainability or collapse: an integrated history and future of people on earth. MIT Press, Cambridge pp 115–148

Santa Fe Institute (2008) Santa Fe Institute research topics. Santa Fe Institute, Santa Fe, New Mexico. www.santafe.edu. 12 Jan 2009

United States Department of Agriculture (2009) National Center for Genetics Resources Preservation (NCGRP) preserving our future. National center for genetic resources preservation. United States Department of Agriculture, Fort Collins, Colorado. http://www.ars.usda.gov/main/site_main.htm?modecode=54020500. Accessed 10 Feb 2009

University of Wisconsin (2008) UW-Madison cluster hiring initiative. http://www.clusters.wisc.edu. 10 Jan 2010

Chapter 8
Monitoring and Assessment—What to Measure … and Why

Charles A. Cole and Mary E. Kentula

Abstract It is often difficult to know what to measure when conducting a wetland assessment. There are a wide variety of variables to consider across the three main parameters—water, vegetation, and soils. To complicate things even further, the aspect of time and space must be considered if you wish to be able to make sense of your assessment. We present a discussion of water, vegetation, and soils and then give our best judgment of which variables to measure for each, and why some might be more useful than others. The merits and problems with gathering data from single visits versus multiple visits are discussed, as well as the level of expertise needed in some instances for certain variables to be useful. We do not discuss assessment and inventory methods as these will follow once you chose your assessment variables most relevant to your goals.

Monitoring is often viewed as the leftovers after the main meal. Researchers love to feast on the investigation at hand, but the leftover issues that require monitoring receive much less attention. There are innumerable examples of wetland research that examine important questions, but then ignore the temporal aspect of these studies that require some form of monitoring. The literature abounds with studies that substitute "space for time," but rarely do we see a work product that actually examines the time component. For example, Ruiz-Jaen and Aide (2005) looked at how wetland restoration success was being measured and found that the element of time through repeated monitoring was rarely addressed. Local and regional hydrology is dynamic and changes through time, especially with drought cycles, yet we rarely monitor site hydrology for more than the time required by a permit or degree program or research grant. Plant communities change through time by means of succession, which is often directly influenced by disturbance, but we typically measure the characteristics of the plant community for only a few years at most.

C. A. Cole (✉)
Department of Landscape Architecture, Penn State University,
121 Stuckeman Family Building, University Park, PA 16802, USA
e-mail: cac13@psu.edu

B. A. LePage (ed.), *Wetlands,*
DOI 10.1007/978-94-007-0551-7_8, © Springer Science+Business Media B.V. 2011

Soils, perhaps the most stable of the "big three" (i.e., vegetation, hydrology, and soil) wetland indicators, still vary through time and space, especially as affected by disturbance, but again, repeated measurements (i.e., monitoring) are rare. Without a doubt, time is a critical element in our understanding of wetland ecology and one that cannot be fully understood using the "space for time" approach. Monitoring is the key to understanding changes in wetlands through time and for knowing how the "space for time" approach can be used most effectively.

We will not discuss assessment and inventory methods in this chapter as these are readily found throughout the literature and the choice of methods is dependent on the objectives and constraints of the individual monitoring program. Our goal here is to inform wetland managers, ecologists, and scientists about the wetland features that are worthwhile monitoring in any wetland assessment. We have focused on the "big three" wetland indicators as we view them to be central to any monitoring program. Additional classes of indicators, such as measures of stressors, algae, wildlife use, and water chemistry, should be considered in light of the objectives of the monitoring program you are implementing.

Hydrology

The first of the "big three" wetland parameters is hydrology. Without water you do not have a wetland. Hydrology, therefore, is the most important aspect of a wetland assessment, but it is also perhaps the most difficult aspect of a wetland to accurately depict and understand. There are a considerable number of studies that have examined the hydrology of specific wetlands in detail (Winter and Rosenberry 1998; Sun et al. 2006; Todd et al. 2006). There are considerably fewer studies that have hydrologic records extending longer than 5 years (Carr et al. 2006; Boswell and Olyphant 2007; Large et al. 2007). Most studies do not examine the hydrology of a large number of wetlands over the span of many years (Cole et al. 2008). In any event, any reasonable characterization of site hydrology is difficult to achieve when there exists a time constraint for assessment. In the United States this is particularly true of jurisdictional wetland determinations (actions where the law specifies that a wetland boundary must be determined on the land). This leaves us with the delicate question of how to assess hydrology over the short term and still tie it to the site's long-term hydrologic regime, both temporally and spatially (Fig. 8.1).

Indicators are frequently used to make determinations of probable wetland hydrology at a site. In the United States, investigators have relied upon several approaches, including, most frequently, those that are based on soils (United States Department of Agriculture and Natural Resources Conservation Service 2006; Richardson et al. 2009), vegetation (Carr et al. 2006; Johnston et al. 2007), and wetland classification (Cole et al. 2008).

Indicators however, do not provide quantitative information on the length of time a wetland is wet or on the dynamics of the water table on the site. Some work on

Fig. 8.1 Downloading water-level data from a recorder

indicators has begun to illuminate these issues (Cole et al. 2008; Richardson et al. 2009), but these studies remain the exception rather than the rule. So the choice of how one assesses hydrology comes down to why the information is needed. Shaffer et al. (2000) discussed the information that one can obtain from different measurement intervals of water level. Infrequently collected data may be perfectly acceptable. For example, sampling wells at monthly intervals, supplemented with crest gages provides a representative hydrograph for a site (Shaffer et al. 2000). The bottom line is that each approach to measuring hydrology (e.g., indicators, long-term depth data) has inherent strengths and weaknesses that need to be acknowledged and considered in light of the study's or project's objectives.

Single Visit Hydrology

Some wetland assessments only allow for a single visit to a site to assess hydrology. In these instances, one must rely upon wetland indicators, despite the limitations of such data. Indicators are, by design, intended to be quick, providing a snapshot of the site's hydrologic conditions. The standard indicators used for jurisdictional assessment in the United States reflect that philosophy: "...drainage patterns, drift lines, sediment deposition, watermarks, stream gage data and flood predictions, historic records, visual observation of saturated soils, and visual observation of inundation" (United States Army Corps of Engineers 1987). There is value in each of these measurements especially if all you need to know is whether a site gets sufficiently wet to support wetland plants. However, none of the indicators mentioned above, with the possible exception of stream gage or historic data, if it exists, will tell you when the hydrologic event(s) occurred, which is problematic when you need to know what actually happened at the site. Furthermore, none of these indicators tells you how often or for how long the site was inundated or saturated, or if there were dry periods interspersed between wet periods.

Of the indicators listed above, drainage pattern may ultimately prove the most useful, if you can correctly classify a wetland system. For example, Cole et al. (1997, 2008) described characteristic hydrologic behavior by wetland type within the Ridge and Valley physiographic (broad-scale subdivisions of the Earth's surface based on terrain texture, rock type, geologic structure, and history) province of Pennsylvania. Certain basic information, such as percent of time that water is located in the root zone, is relatively predictable if you can correctly classify certain types of groundwater systems. That indicator however, is less predictable with surface water wetlands, such as flood plain systems. For single visits, knowledge of the wetland type is crucial, and that knowledge should be tied to the hydrologic data from a regional reference wetland system. Looking for alterations to natural hydrology (e.g., ditches) on a site is also important, at least to identify disturbance to a site. Such information cannot tell you where the water was located necessarily, but these data will give you some indication that change, often negative, has occurred.

For monitoring hydrology, single visits are problematic for the simple reason that you never know where you will "land" on the annual hydrograph. Single visit monitoring is especially difficult with created and restored wetlands. Some of the indicators listed above may not have had time to develop and, therefore, would not be reliable. However, many created wetlands are built around permanent water (ponds) and thus are readily assessed in the sense of having too much water too much of the time (Shaffer et al. 1999; Cole and Brooks 2000).

Multiple Visit Hydrology

Multiple visits to a wetland to assess hydrology are always preferred so as to get some idea of seasonal water levels (frequency) and spatial dynamics (variability in water levels and flow). It is not always necessary to electronically instrument a wetland to gain an in-depth understanding of its hydrologic dynamics—sometimes simple slotted pipe wells will suffice (Fig. 8.2). Again, how often you take measurements, and where you place any instruments, depends upon the question(s) to be answered.

The frequency of measurement of hydrology in wetlands was addressed by Shaffer et al. (2000). The results from sampling a combination of a well and crest gage (Fig. 8.2) at monthly intervals was roughly comparable to sampling at daily intervals when estimating the median and range of the water levels. Dynamics of water movement in and out of the root zone were less well addressed. Such data are important when asking what percentage of time water is present in the root zone. For those who need exceedance data (i.e., when water was in the root zone or not), sampling every two days (or more often) is necessary. Ordoyne and Friedl (2008) found a 16-day sample interval to be appropriate for assessing wetland hydroperiod in the Florida Everglades using remote sensing techniques.

The frequency of measurements should also include some discussion of how long the hydrologic data are to be collected. Management actions will often dictate

Fig. 8.2 Example of
site equipped for general
hydrologic monitoring as
described by Shaffer et al.
(2000). The well is used for
measuring water levels below
the surface of the substrate,
while the staff gage is used
to measure water levels
above the surface of the
substrate. The crest gage is
used to determine the highest
water level that occurred
during the interval between
measurements

a short time interval of measurement before some action occurs (e.g., a permitted impact). Sampling over a short time period can lead to spurious and potentially misleading data, especially if the data are collected in either very wet or very dry years. A glance at the hydrograph in Fig. 8.3 shows the possibility of error if the data were collected during these times.

Monitoring wetland hydrology can appear to be a never-ending project and we do not advocate open-ended monitoring. However, we also do not advocate single visit or short-term monitoring. If you need to monitor a wetland to assess some regulatory compliance issue (such as delineating a problematic wetland site), then multiple years should be the norm, even if only at a monthly interval. If you are monitoring wetlands to assess the success of a restoration project, then detailed monitoring is needed for the time that it takes for the site to get established to ensure that the correct hydrology has been established. This will be a number of years, depending on the wetland type and whether corrections need to be made in the

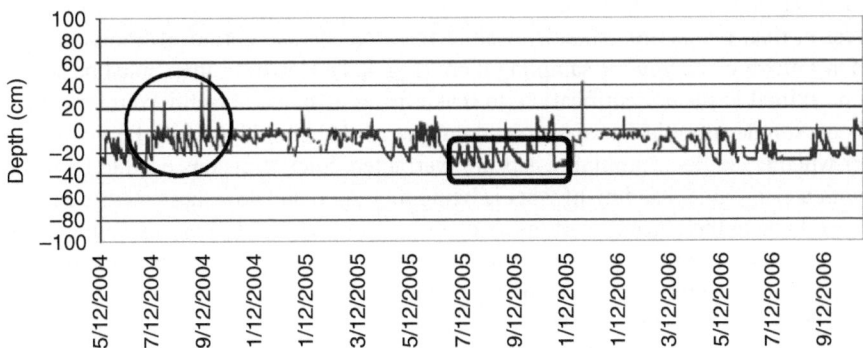

Fig. 8.3 Long-term hydrologic monitoring can help to eliminate errors such as those that might occur if you only sampled within the dates included in the *circle* (wetter) or in the *rectangle* (drier)

water regime. Getting the hydrology correct is the most important aspect of wetland restoration (Hoeltje and Cole 2007).

Spatial Aspects

Assessing wetland hydrology on a site classically involves the installation of a large network of shallow groundwater wells and piezometers (Winter and Rosenberry 1998). For most projects involving the monitoring of hydrology, such networks are not feasible and a less intensive approach is taken. To a large extent, we rely on our best professional judgment (BPJ) when it comes to the placement of water level recording wells. Cole et al. (1997) placed recorders across the visible hydrologic gradient. Magee and Kentula (2005) extrapolated water depths based upon permanent monitoring stations set at the lowest point within a wetland. Large et al. (2007) located recorders near areas of likely recharge and discharge and Sun et al. (2006) used a large number of wells and piezometers across many directions in an intensive study of a Carolina Bay. Some sites, such as groundwater-fed wetlands, may be assessed using only a single well (see Shaffer et al. 1999 for an example of such a study), but others, such as floodplain wetlands, may require many wells to obtain adequate information. There is no clear answer on the best methodology or density of monitoring equipment, as the data quality required to meet the objectives of the study or project (as well as, time and resources) will dictate what you are able to do. In general, however, the more wells and/or piezometers you have installed, the more complete the picture of the wetland's hydrology will be.

Vegetation

The second of the "big three" wetland parameters is the vegetation. Monitoring wetland vegetation is a somewhat simpler task than measuring hydrology, if for no other reason than plants are generally visible for a reasonable period of time and they don't move while you are sampling them (Fig. 8.4). However, there must be someone skilled in plant identification, to make the assessment a useful exercise. There is no simple formula for what it is about plants that should actually be monitored or whether single or multiple visits are warranted. Such decisions need to be made with a clear understanding of what is being measured and how these measurements are related to the objectives of the study or project. Cole (2002) discussed the problems with the common practice of relying on the percent plant cover as the only or the major indicator when monitoring wetlands. He noted very little correlation between plant cover and any actual wetland function. As with most monitoring efforts, however, the cost and logistic constraints and, in some cases, legal matters in the United States often force quick visits to wetlands for assessment purposes leaving us with the same conundrum—what useful information can we get during these limited monitoring efforts and can they replace long-term, comprehensive monitoring?

Fig. 8.4 Sampling vegetation in a 1-meter square quadrat in a riverine wetland

Single Visit Botany

Generalizing about monitoring plants is difficult because so much depends upon geography and the prevailing climate. Monitoring herbaceous plants in Florida, for example, can occur year round, whereas monitoring in Pennsylvania really only happens between June and September (although skunk cabbage [*Symplocarpus foetidus*] emerges in February leaving us with a bit of a philosophical dilemma with regards to the definition of the "growing season", no?). To some degree however, the time of year may be irrelevant if you are conducting a floristic quality assessment and using coefficients of conservatism (C). Swink and Wilhelm (1994) developed C as a means of assigning plant species to a relatively unaltered landscape versus one that has been more heavily impacted. The frequency or the more often a plant species is found in the unaltered landscape, the higher the C value. Conversely, a low C results from a plant species being more commonly located in disturbed settings. After C values are assigned to the flora of an area they can be combined into a Floristic Quality Index (FQI) that can be used to assess the ecological condition of wetlands in the region (Lopez and Fennessy 2002). Interestingly, while the spring and summer flora on a site may be very different, the FQI from each community may well not differ. If the FQI is indeed an inherent characteristic of the flora on a site, then the FQI on a site should reflect the same general value regardless of time of year. Thus, you might be able to sample when you are able, as opposed to sampling when a certain segment of the flora is in flower.

Although the concepts of C and the FQI seem to have high potential for monitoring vegetation and using the data to infer the degree of human disturbance, not every region has C values available or FQI's calculated (Ervin et al. 2006). Therefore, other aspects of the plant community must be considered for use in monitoring. If we move past the typically used "percent herbaceous cover", as suggested by Cole (2002), then what is left? An especially useful parameter, for example, is the percent of invasive and exotic (i.e., non-native) plant species in the floral assemblage. Some attributes, such as the percent plant cover and biomass may provide particularly useful information when they are tied to additional data such as biomass by species or vegetation strata. Also useful can be diversity indices (e.g., Shannon-Weiner), number and type of strata, and interspersion of different plant communities on a site.

Mack and Kentula (2010) evaluated 40 existing vegetation indices of biological condition (IBI) developed for wetlands in an effort to identify convergence in the type of vegetation attributes that are ultimately selected as metrics. An attribute is a measureable feature of the group being evaluated; a metric is an attribute selected for inclusion in an assessment method like an IBI (Karr and Chu 1999). The premise behind the effort was that method convergence might indicate the existence of core metrics applicable to multiple regions and to different wetland types. Mack and Kentula found that the metrics composing the 40 indices sorted into 24 categories of which over 40% were used by more than 50% of the indices. Measures of invasive species were used by all the indices. The ten most commonly used categories, in rank order, were: invasive species metrics; sensitive species metrics; annual/biennial/perennial metrics; total taxa metrics; tolerant species metrics; FQI metrics; native graminoid metrics; hydrophytic metrics; aquatic guild metrics; and invasive graminoid metrics.

Single visits are clearly not intended to assess change, but can be useful to characterize sites of differing age or under different disturbance regimes. Under these circumstances, simple plant species lists are very useful, especially if you can classify the plants into categories such as functional groups (Bouton and Keddy 1993; Reich et al. 2004; Petchey and Gaston 2006). In some early work on the application of this concept to wetlands, Menges and Waller (1983) used the three functional groups (competitive, stress tolerant, and ruderal) recognized by Grime (1977, 1979) to describe wetland plants growing along an elevation gradient on a floodplain.

Information on which functional group a plant belongs to is not always available. However, the United States Fish and Wildlife Service (USFWS) has classified almost all wetland plants as to their wetland indicator status (WIS) (Reed 1988, as modified in 1996 [United States Army Corps of Engineers 1996]) and the United States Army Corps of Engineers (ACOE) is in the process of developing regionalized lists of hydrophytic plants. The list is comprehensive and widely used. Thus, this functional classification allows you, at a minimum, to understand whether the flora is typical of wetlands or one that is more commonly found in uplands (though it does not reflect important characteristics such as whether invasive plant species dominate). In an application of this concept, Wentworth et al. (1988) showed how to use the WIS to develop indices for determining if an area meets the legal definition of a wetland in the United States.

Multiple Visit Botany

The advantages to multiple visits to a site are obvious—with enough visits over time, you can begin to detect floristic change. We are not necessarily advocating decadal measurements, though such efforts are admirable and produce tremendous information (Skirvin et al. 2008). Bestelmeyer et al. (2006) detected dynamic patterns on the scale of 20–40 years. However, it seems intuitive at some level to suggest measurements every 2–3 years in areas that are under some pressure or undergoing change, like the establishment of a wetland after restoration. The woody vegetation is unlikely to change much over this length of time (barring any major disturbances) and the interval is short enough that you can still catch relatively quick changes in the herbaceous layer (e.g., from invasive species [Romanello 2009]).

If you have the resources, then you can monitor vegetation communities using remote sensing. Laba et al. (2008) for example, generated map accuracies of up to 83% for some invasive plant species. Lee and Yeh (2009) were able to discern changes in mangroves over a 10-year span using satellite imagery, thus developing a hybrid single-multiple visit approach. Such success depends upon having the imagery over a time series and the ability to interpret it correctly. As for the frequency of such efforts, Lunetta et al. (2004) suggest a minimum of every 3–4 years to be able to accurately assess land cover change. This is problematic for some areas where aerial imagery may not be available at that frequency.

Many of the attributes useful for single visits such as C, species lists, and WIS, are applicable for multiple visits as well. Any aspect of the plant community that is measured repeatedly is potentially useful for assessing change.

It is worth remembering however, that the parameters described above generally measure community structure and not wetland function. It is rare that one actually measures rates and processes, the true definition of function (Simenstad and Thom 1996; Hruby 1999). There is nothing wrong with measuring structure and its change over time, but we need to be clear as to what is being monitored and what is not. Furthermore, monitoring should occur both before and after any impact if you want to look for cause and effect. If monitoring is performed ahead of an impact, then the better your conclusions about any change that has occurred will be.

Soils

Soils are perhaps the simplest to monitor of the three wetland parameters, at least operationally (hereby annoying all of our soil scientist colleagues, we realize). We speak merely in reference to the fact that soils do not move (unless you're in California after a fire or an earthquake) and soils can reflect the past presence of hydric conditions for a very long time, thus increasing their utility as an indicator when a site has been ditched, drained, and stripped of vegetation. For those of whose expertise in soils is limited to "yes, it's mucky", the language of soil science can be a bit obtuse. Nonetheless, an understanding of hydric soils, their formation, and their behavior over time is absolutely critical to successful monitoring (Fig. 8.5). In particular, a description of the soil profile can confirm the presence of conditions necessary for hydric soil development and function (Figs. 8.6 and 8.7).

Fig. 8.5 Wetland gleyed soil.
(Image provided copyright
free from Jonty68 at the English Wikipedia project)

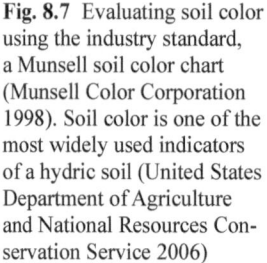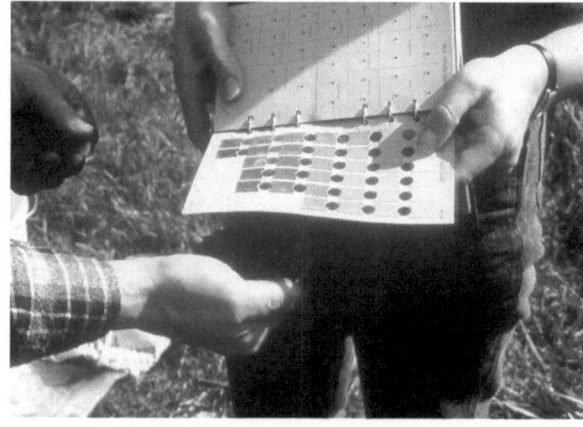

Fig. 8.6 Layers in a soil slab extracted from a pit in a wetland. Such a sample is used to describe
the soil characteristics such as color (Fig. 8.7) and texture. Samples of the soil may also be taken
for chemical analyses such as soil organic matter content

Fig. 8.7 Evaluating soil color
using the industry standard,
a Munsell soil color chart
(Munsell Color Corporation
1998). Soil color is one of the
most widely used indicators
of a hydric soil (United States
Department of Agriculture
and National Resources Conservation Service 2006)

Single Visit Pedology/Edaphology

As with water and plants, the utility of monitoring soils with a single visit depends entirely upon what it is you wish to determine. Bestelmeyer et al. (2006) examined plant dynamics over a period of approximately 40 years, but only measured soil characteristics once, a testament, perhaps, to the perceived stability of soils.

Any number of characteristics can be included in a single visit, such as presence/absence of hydric soils (including hydric soil indicators [United States Department of Agriculture and National Resources Conservation Service 2006]), evidence of site alteration (e.g., was it drained), or nutrient levels. These provide important and useful information, but say little about the temporal aspects that can only be discerned through multiple visits. A number of useful manuals and guides to sampling soils are available, especially through the United States Department of Agriculture, National Resources Conservation Service (USDA-NRCS) (Thien 1979; Schoeneberger et al. 2002; Soil Survey Staff 2006). Also see the websites of the NRCS and ACOE for the most current information.

Multiple Visit Pedology/Edaphology

Monitoring over time may seem unnecessary given the apparent stability of soils, but evidence is beginning to suggest that repeated measurements may be needed. Sparling et al. (2004) developed a soil quality monitoring protocol for New Zealand and recommended resampling every 3–10 years, with impacted sites being sampled more frequently. Changes in the soil flora might even happen more quickly, as in the span of months. Kourtev et al. (2003) found changes in the soil microflora after only three months in sites invaded by Japanese stiltgrass (*Microstegium vimineum*). Wolfe and Klironomos (2005) document other impacts on the soil. Parkin and Kaspar (2004), during March–June 2006, looked at the effects of sampling frequency on CO_2 flux and found little change in the results between 12 and 20 day intervals, with increased precision less than that, and a definite loss afterwards. Brejda et al. (2000) discussed the difficulties in getting usable repeat samples using the USDA-NRCS 5-year sampling protocol (see Kellogg et al. 1994).

What to measure? The parameters given above in the single visit section are still important, and getting temporal data greatly increases their value. The range of choices for other parameters is huge. Ritz et al. (2009) reviewed almost 200 biological characteristics that related to soil functions in an effort to answer that question. They decided that measurements strongly tied to the structure of the soil microbial community were most useful. Garten et al. (2003) looked at both soils and vegetation to examine the impacts of military training on ecosystem structure. For soils, they measured particulate organic matter, mineral-associated organic matter, refractory carbon, and total carbon and nitrogen concentrations. Craft et al. (2007) found that only total soil phosphorous (P) was related to wetland nutrient condition when looking at nutrient-enriched wetlands. In any event, as with water and plants, what you measure might ultimately be limited by budget and time. A useful candidate measure for general use is soil organic matter (SOM) as SOM is tightly tied to many wetland functions (see Shaffer and Ernst 1999 and the references cited therein). Even though an indicator as conceptually simple as SOM might take decades to reach reference levels in created wetlands (Ballantine and Schneider 2009), getting temporal trends might be extremely useful in identifying impact to a site and for tracking progress in the development of restoration projects.

Reference Sites

Monitoring must have some aspect of comparison to increase the value of the information gathered. Documenting the development of a created wetland over five years is useful, but to then compare those data with the same information from a natural wetland of the same type is much more valuable. Herein lays the value of reference wetlands. The development of the notion of reference for wetlands began in earnest with the development of the hydrogeomorphic approach to wetland assessment (Brinson 1993). Rheinhardt et al. (1997) argue that reference sites should define the range of conditions found on natural wetlands in an area (i.e., across a gradient of disturbance) (Fig. 8.8). They refer to the very best, least impacted sites as reference standards. Reference wetlands are now regularly included in many site assessment projects, especially those involved in regulatory actions.

Developing a suite of reference wetlands within a region should be of the highest priority as these sites can then serve a variety of uses, from basic ecologic data to design and monitoring templates for restoration efforts. The effort required to do this is substantial. In Pennsylvania, over 200 reference wetlands have been established (Brooks et al. 2004), but the effort has taken well over 10 years. However, these data have been extremely useful in understanding hydrology (Cole et al. 1997; Cole and Brooks 2000) and assessing the effects of wetland degradation (Brooks

Fig. 8.8 Three floodplain wetlands in central Pennsylvania. Each is under a different level of disturbance: least on the *left*, moderate *center*, and severe on *right*. The decreasing amounts of woody vegetation from *left* to *right* reflects the increasing disturbance levels in these three wetlands

et al. 2005), among other results. If you are located near such a set of reference wetlands, perhaps developed by a state or a university group, then your ability to make meaningful comparisons is greatly increased.

Summary

We hope that we have made it clear that the choices for monitoring wetlands are neither always obvious nor easy to implement. The decisions made relative to monitoring should be based on the objectives of your study or project and the intended use of the data. Those decisions will also be mediated by the level of effort that can be supported by the resources available. Of critical importance, regardless of what choices are made, is to instill an element of time in your monitoring and assessment actions. While this might seem obvious, long-term monitoring and assessment rarely occurs and we lose valuable information that might well be critical to the management of some of our more important aquatic resources.

We also hope that we have made it clear that there are huge pay-offs for monitoring, especially monitoring that is well planned and executed. Monitoring data are essential to documenting the condition of the wetland resource, for determining the success of management actions such as restoration, and for understanding how wetland ecosystems change over time in the face of human-induced disturbances.

References

Ballantine K, Schneider R (2009) Fifty-five years of soil development in restored freshwater depressional wetlands. Ecol Appl 19:1467–1480

Bestelmeyer BT, Trujillo DA, Tugel AJ, Havstad KM (2006) A multi-scale classification of vegetation dynamics in arid lands: what is the right scale for models, monitoring, and restoration? J Arid Environ 65:296–318

Boswell JS, Olyphant GA (2007) Modeling the hydrologic response of groundwater dominated wetlands to transient boundary conditions: implications for wetland restoration. J Hydrol 332:467–476

Boutin C, Keddy PA (1993) A functional classification of wetland plants. J Veg Sci 4:591–600

Brejda JJ, Moorman TB, Smith JL, Karlen DL, Allan DL, Dao TH (2000) Distribution and variability of surface soil properties at a regional scale. Soil Sci Soc Am J 64:974–982

Brinson MM (1993) A hydrogeomorphic classification for wetlands. United States Army Corps of Engineers, Technical Report WRP—DE—4. Washington, DC, p 101. http://el.erdc.usace.army.mil/wetlands/wlpubs.html

Brooks RP, Wardrop DH, Rubbo JM, Mahaney WM, Cole CA (2004) Hydrogeomorphic functional assessment models by wetland type for Pennsylvania ecoregions. Section II.B.3.b.4. In: Brooks RP (ed) Monitoring and assessing Pennsylvania wetlands. Final Report for Cooperative Agreement No. X-827157-01, between the Penn State Cooperative Wetlands Center, Pennsylvania State University, University Park, PA and U.S. EPA, Office of Wetlands, Oceans, and Watersheds, Washington, DC

Brooks RP, Wardrop DH, Cole CA, Campbell DA (2005) Are we purveyors of wetland homogeneity?: A model of degradation and restoration to improve wetland mitigation performance. Ecol Eng 24:331–340

Carr DW, Leeper DA, Rochow TF (2006) Comparison of six biologic indicators of hydrology and the landward extent of hydric soils in west-central Florida, USA cypress domes. Wetlands 26:1012–1019

Cole CA (2002) The assessment of herbaceous plant cover in wetlands as an indicator of function. Ecol Indic 56:1–7

Cole CA, Brooks RP (2000) A comparison of the hydrologic characteristics of natural and created mainstem floodplain wetlands in Pennsylvania. Ecol Eng 14:221–231

Cole CA, Brooks RP, Wardrop DH (1997) Wetland hydrology and water quality as a function of hydrogeomorphic subclass. Wetlands 17:456–467

Cole CA, Cirmo CP, Wardrop DH, Brooks RP, Peterson-Smith JE (2008) Transferability of an HGM wetland classification scheme to a longitudinal gradient of the central Appalachian Mountains: initial hydrological results. Wetlands 28:439–449

Craft C, Krukk K, Graham S (2007) Ecological indicators of nutrient enrichment, freshwater wetlands, midwestern United States (U.S.). Ecol Indic 7:733–750

Ervin GN, Herman BD, Bried JT, Holly DC (2006) Evaluating non-native species and wetland indicator status as components of wetlands floristic assessment. Wetlands 26:1114–1129

Garten CT, Ashwood TL, Dale VH (2003) Effect of military training on indicators of soil quality at Fort Benning, Georgia. Ecol Indic 3:171–180

Grime JP (1977) Evidence of the existence of three primary strategies in plants and its relevance to ecological and evolutionary theory. Am Nat 111:1169–1194

Grime JP (1979) Plant strategies and vegetation processes. Wiley, Chichester, p 222

Hoeltje SM, Cole CA (2007) Losing function through wetland mitigation in central Pennsylvania, USA. Environ Manag 39:385–402

Hruby T (1999) Assessment of wetland functions: what they are and what they are not. Environ Manag 23:75–85

Johnston CA, Bedford BL, Bourdaghs M, Brown T, Frieswyk C, Tulbure M, Vaccaro L, Zedler JB (2007) Plant species indicators of physical environment in Great Lakes coastal wetlands. J Great Lakes Res 33:106–124

Karr JR, Chu EW (1999) Restoring life in running waters: better biological monitoring. Island Press, Washington, p 220

Kellogg RL, TeSelle GW, Goebel JJ (1994) Highlights from the 1992 National Resources Inventory. J Soil Water Conserv 49:521–527

Kourtev PS, Ehrenfeld JG, Haggblom M (2003) Experimental analysis of the effect of exotic and native plant species on the structure and function of soil microbial communities. Soil Biol Biochem 35:895–905

Laba M, Downs R, Smith S, Welsh S, Neider C, White S, Richmond M, Philpot W, Baveye P (2008) Mapping invasive wetland plants in the Hudson River National Estuarine Research Reserve using quickbird satellite imagery. Remote Sens Environ 112:286–300

Large ARG, Mayes WM, Newson MD, Parkin G (2007) Using long-term monitoring of fen hydrology and vegetation to underpin wetland restoration strategies. Appl Veg Sci 10:417–428

Lee T, Yeh H (2009) Applying remote sensing techniques to monitor shifting wetland vegetation: a case study of Danshui River Esturary mangrove communities, Taiwan. Ecol Eng 35:487–496

Lopez RD, Fennessy MS (2002) Testing the floristic quality assessment index as an indicator of wetland condition. Ecol Appl 12:487–497

Lunetta RS, Johnson DM, Lyon JG, Crotwell J (2004) Impacts of imagery temporal frequency on land-cover change detection monitoring. Remote Sens Environ 89:444–454

Mack JJ, Kentula ME (2010) Metric similarity in vegetation-based wetland assessment methods. EPA/600/R-10/140. United States Environmental Protection Agency, Office of Research, Office of Research and Development, Washington, DC

Magee TK, Kentula ME (2005) Response of wetland plant species to hydrologic conditions. Wetlands Ecol Manag 13:163–181

Menges ES, Waller DM (1983) Plant strategies in relation to elevation and light in flood plain herbs. Am Nat 122:454–473

Munsell Color Corporation (1998) Munsell soil color charts. Gretag Macbeth, New Windsor

Ordoyne C, Friedl MA (2008) Using MODIS data to characterize seasonal inundation patterns in the Florida Everglades. Remote Sens Environ 112:4107–4119

Parkin TB, Kaspar TC (2004) Temporal variability of soil carbon dioxide flux: effect of sampling frequency on cumulative carbon loss estimation. Soil Sci Soc Am J 68:1234–1241

Petchey OL, Gaston KJ (2006) Functional diversity: back to basics and looking forward. Ecol Lett 9:741–758

Reed PB Jr (1988) National list of plant species that occur in wetlands: national summary. United States Fish and Wildlife Service, Biological Report 88, pp 1–244

Reich PB, Tilman D, Naeem S, Ellsworth DS, Knops J, Craine J, Wedin D, Trost J (2004) Species and functional group diversity independently influence biomass accumulation and its response to CO_2 and N. Proc Natl Acad Sci 101:10101–10106

Rheinhardt RM, Brinson M, Farley PM (1997) Applying wetland reference data to functional assessment, mitigation and restoration. Wetlands 17:195–215

Richardson TC, Robison CP, Neubauer CP, Hall GB (2009) Hydrologic signature analysis of select organic hydric soil indicators in northeastern Florida. Soil Sci Soc Am J 73:831–840

Ritz K, Black HIJ, Campbell CD, Harris JA, Wood C (2009) Selecting biological indicators for monitoring soils: a framework for balancing scientific and technical opinion to assist policy development. Ecol Indic 9:1212–1221

Romanello GA (2009) *Microstegium vimineum* invasion in central Pennsylvanian slope, seep wetlands: site comparisons, seed bank investigation and water as a vector for dispersal. Master's Dissertation, Penn State University, University Park, Pennsylvania, p 104

Ruiz-Jaen MC, Aide TM (2005) Restoration success: how is it being measured? Restor Ecol 3:569–577

Schoenberger PJ, Wysocki DA, Benham EC, Broderson WD (2002) Field book for describing and sampling soils, Version 2.0. Natural Resources Conservation Service, National Soil Survey Center, Lincoln, Nebraska, p 228

Shaffer PW, Ernst TL (1999) Distribution of soil organic matter in freshwater emergent/open water wetlands in the Portland, Oregon metropolitan area. Wetlands 19:505–516

Shaffer PW, Kentula ME, Gwin SE (1999) Characterization of wetland hydrology using hydrogeomorphic classification. Wetlands 19:490–504

Shaffer PW, Cole CA, Kentula ME, Brooks RP (2000) Effects of measurement frequency on water level summary statistics. Wetlands 20:148–161

Simenstad CA, Thom RM (1996) Functional equivalency trajectories of the restored Gog-Le-Hi-Te estuarine wetland. Ecol Appl 6:38–56

Skirvin S, Kidwell M, Biedenbender S, Henley JP, King D, Collins CH, Moran S, Weltz M (2008) Vegetation data, Walnut Gulch Experimental Watershed, Arizona, United States. Water Resour Res 44:W05S08. doi:10.1029/2006WR005724

Soil Survey Staff (2006) Keys to soil taxonomy. United States Department of Agriculture, Natural Resources Conservation Service, Washington, p 341

Sparling GP, Schipper LA, Bettjeman W, Hill R (2004) Soil quality monitoring in New Zealand: practical lessons from a 6-year trial. Agric Ecosyst Environ 104:523–534

Sun G, Callahan TJ, Pyzoha JE, Trettin CC (2006) Modeling the climatic and subsurface stratigraphy controls on the hydrology of a Carolina Bay wetland in South Carolina, USA. Wetlands 26:567–580

Swink F, Wilhelm G (1994) Plants of the Chicago region, 4th edn. Indiana Academy of Science, Indianapolis, p 921

Thien SJ (1979) A flow diagram for teaching texture by feel analysis. J Agron Educ 8:54–55

Todd AK, Buttle JM, Taylor CH (2006) Hydrologic dynamics and linkages in a wetland-dominated basin. J Hydrol 319:15–35

United States Army Corps of Engineers (1987) Corps of Engineers wetlands delineation manual. Technical Report Y-87-1. United States Army Engineer Waterways Experiment Station, 3909 Halls Ferry Road, Vicksburg, Mississippi, p 143

United States Army Corps of Engineers (1996) National list of vascular plant species that occur in wetlands: 1996 national summary. p 209. http://www.usace.army.mil/CECW/Documents/cecwo/reg/plants/national.pdf

United States Department of Agriculture and Natural Resources Conservation Service (2006)
    Field Indicators of hydric soils in the United States: guide for identifying and delineating hy-
    dric soils, Version 6.0. Natural Resources Conservation Service, Washington, DC, p 39
Wentworth TG, Johnson GP, Kologiski RL (1988) Designation of wetlands by weighted averages
    of vegetation data: a preliminary approach. Water Resour Bull 24:389–396
Winter TC, Rosenberry DO (1998) Hydrology of prairie pothole wetlands during drought and
    deluge: a 17-year study of the Cottonwood Lake wetland complex in North Dakota in the per-
    spective of longer term measured and proxy hydrological records. Clim Change 40:189–209
Wolfe BE, Klironomos JN (2005) Breaking new ground: soil communities and exotic plant inva-
    sion. BioScience 55:477–487

Part III
The Anthropogenic Framework

Chapter 9
Wetlands from a Psychological Perspective: Acknowledging and Benefiting from Multiple Realities

Rachel Kaplan

Abstract Information sharing is a necessity among collaborating team members as well as with many constituencies. It is easy to take it for granted, until it fails to work well. The chapter is largely devoted to what leads to more effective and less frustrating information sharing. Rather than a long list of do's and don'ts—which are not a very effective way to share information—we use a simple framework that makes it easier to think about the issues. The same framework turns out to be useful for discussing the role that wetlands play in the urban context. For the most part urban residents have little understanding of the experts' criteria for successful wetlands efforts. To them the wetland is "nature" and nature matters. The chapter thus also includes discussion of the importance of nature to the human consumers of the wetland project. The framework that brings all this together, the Reasonable Person Model (RPM), focuses on how environments or situations people find themselves in can be more supportive of their effectiveness and reasonableness.

Even the best intentions and years of training are often insufficient for many of the challenges that get in the way of doing our jobs. Wetland restoration may often seem far more straightforward and manageable to accomplish were it not for the people one has to deal with. It is not just the obstreperous people, the ones who seem opposed to just about anything. Nor is it just about the ignorance so many people have about the science and technology that are involved. Even some members of a team working on a project can be frustrating despite their training and shared appreciation for the goals of the project.

There are meetings and discussions and more meetings and then it still seems that everyone is not on the same page. Despite agreeing on endless details, there are still surprises, unanticipated consequences, and regrets. Hindsight is enriching, but

R. Kaplan (✉)
School of Natural Resources and Environment, University of Michigan,
440 Church Street, Ann Arbor, MI 48109-1041, USA
e-mail: rkaplan@umich.edu

B. A. LePage (ed.), *Wetlands,*
DOI 10.1007/978-94-007-0551-7_9, © Springer Science+Business Media B.V. 2011

foresight would seem to be more satisfying. Why are these patterns so common? Are there ways to make them less likely?

When thinking about criteria for assessing the success of a wetland project the list is likely to be long. No doubt it would reflect the many considerations raised by regulatory agencies as well as the concerns of each of the specialists on the project team. Despite its length, however, the criteria discussed in this chapter may not make it to the list. Many of these, however, have a great deal to say about the project's success and the process of getting there. In particular, the chapter examines two kinds of concerns that are likely to relate to the success of a project: **information sharing** and the **role of nearby nature in human well-being**.

Information sharing is a necessity among collaborating team members as well as with many constituencies. It is easy to take it for granted, until it fails to work well. This chapter is largely devoted to what leads to more effective and less frustrating information sharing. Rather than a long list of do's and don'ts—which are not a very effective way to share information—we use a simple framework that makes it easier to think about the issues.

The same framework turns out to be useful for discussing the role that wetlands play in the urban context. For the most part urban residents have little understanding of the experts' criteria for successful wetland restoration, enhancement, or creation. To them a wetland is "nature" and nature matters. This chapter thus includes discussion of the importance of nature to the human consumers of a wetland project.

Clearly then, this chapter first needs to talk about the promised framework. It's called the Reasonable Person Model (RPM) because it focuses on what leads people to act more reasonably. All of us have the potential of being unreasonable some of the time. What triggers the unreasonable behavior most often is situational—the contexts, settings, and circumstances that people find themselves in. We like to say that RPM is about how the environment can bring out the best in people (Kaplan and Kaplan 2005, 2009). The framework thus examines how the environments people find themselves in can be more supportive of their effectiveness, and reasonableness.

The Reasonable Person Model (RPM)

At a very young age children start asking "why" about anything and everything. While the "why" questions may be less explicit later on, they remain a persistent quest, about very small matters ("Why don't they say how long this detour in the road will take?") as well as bigger ones ("Why me?"). "Why" is, among other things, an expression of a great desire to understand or make sense of things as well as great interest in exploring the world around us. In a sense, it serves as a powerful indicator of the extent to which humans are information-based creatures. People seek information endlessly, and they get overwhelmed by it. They treasure it, yet resist it. It is the source of truths and lies, insights, and insults. Information is not limited to the spoken word. When a group of people view a wetland they gain a great variety of information from it, although that information may have little in common. The environment is an endless source of information.

RPM focuses on three major domains related to this information-dependence. While highly interconnected and often interdependent, each domain addresses different qualities that contribute to fostering reasonableness. The first domain, **model building**, is closest to the "why" question; it focuses on people's need to understand what is going on around them, which is central to building the mental models that guide our lives. The second domain, **being effective**, looks at the capacity to utilize knowledge and skill as well as ways to offset the pervasive negative impacts of too much information. The third domain, **meaningful action**, relates to the human desire to be needed, to make a difference, and to be treated respectfully.

Model Building

Even without laptops or access to the Internet humans always carry with them enormous amounts of information. The information enables prediction of what might happen next and evaluation of potential consequences of alternative courses of action. It influences perception of what is going on and guides action. This vast amount of information is stored in mental models—simplified versions of reality that are used to make sense of things, to plan, and to evaluate possibilities. The capacity to construct and use such mental models has been (and continues to be) a key tool in human survival.

The dependence on mental models for human functioning would further suggest that people care deeply about information that supports creating and sustaining them. It is thus hardly surprising that people crave information, desire to understand what is going on, and are discomforted by confusion. The motivation to understand also inspires people to venture forth to seek further information. In fact, humans can be quite restless in the absence of such opportunities and pained when their path to exploration is obstructed. Thus in RPM the dual vectors of understanding and exploration are the key components of model building (Kaplan 1973).

Being Effective

The attraction to information can also be debilitating and contribute to a reduced sense of being effective. "Stressed out" has become a common expression for the many ways people feel overwhelmed by opportunities, demands, and responsibilities. However, "stressed out" is not simply the result of too much information. The presence of information *per se* need not be overwhelming; in fact, sitting in a library can be quite peaceful despite the information-rich setting. What leads to the negative state thus is not the presence of information, but the need to pay attention to it. Attention, as Simon (1971) recognized long ago, must be recognized as a scarce resource. The debilitating consequences of information are, then, the result of over-using one's attentional capacity.

Focusing on attention is pivotal to making RPM a helpful tool. With respect to information sharing (this chapter's first theme) it is the attentional capacities of all potential recipients of the information that must be considered. However, when focusing on wetlands as part of people's nearby natural environment (this chapter's second theme), a different kind of attention comes into play. Attention Restoration Theory (ART) (Kaplan 1995, 2001) examines the fascinating interplay between these two dimensions of attention. ART explains how mental fatigue is the result of the depleted capacity to direct one's attention. The cumulative effect of the effort entailed in directing attention can lead to such symptoms as irritability, impatience, and distractibility. ART also points to ways to recover from the fatigued mental state. This involves settings and activities that, rather than requiring that one pay attention, draw on an effortless kind of attention. This kind of attention comes in many forms and at varying scales. For example, one can be mentally exhausted, but have great patience watching the antics of a squirrel, or being engrossed in a novel, or taking a neighborhood walk. Taking a break from expending one's limited capacity to direct attention by doing things that have fascination can thus help one recover from mental fatigue.

Meaningful Action

Information can be enriching and wondrous. It can also lead to a colossal sense of futility and helplessness—hardly states of mind that would lead to reasonableness. By contrast, opportunities for exercising one's effectiveness serve as important examples of meaningful action. Such opportunities can involve even simple steps like having one's opinion sought. The sense that one is making a difference can go a long way toward bringing out the best in people.

The desire to make a difference is exemplified by a wide range of human actions. In the present context, a particularly pertinent example is the thousands of individuals who volunteer their time in restoration and other environmental stewardship efforts. Worldwide, individuals of all ages and backgrounds are dedicating themselves to improving their local environmental conditions through participation in non-govern-mental organizations (NGOs). Consider the subtitle of Hawken's (2007) book: *How the Largest Movement in the World Came into Being and Why No One Saw it Coming.* The book and its related online resource (www.wiserearth.org) provide encyclopedic documentation of organizations "dedicated to restoring the environment and fostering social justice." Hawken discusses how these groups "collectively comprise the largest movement on Earth, a movement that has no name, leader, or location, and that has gone largely ignored by politicians and the media. Like nature itself, it is organizing from the bottom up, in every city, town, and culture and is emerging to be an extraordinary and creative expression of people's needs worldwide." It is certainly a vivid expression of the human propensity for making a difference.

Fig. 9.1 Reasonable Person
Model (RPM)

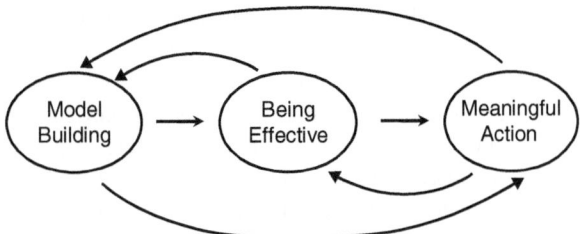

Opportunities for meaningful action are, of course, not limited to volunteer efforts. Unfortunately, many work situations undermine this vital concern. How often do employees feel that their views are not sought or respected, that their efforts to make a difference are stifled, that their jobs lack a sense of meaningfulness? Like so many issues related to RPM, small steps can often make big differences in promoting reasonableness.

Figure 9.1 emphasizes that the three domains of RPM are strongly interrelated. For example, it is easier to engage in stewardship activities if one has some understanding of what it entails. Efforts to make a difference can be undermined by being overwhelmed with information that is presented in ways that make model building difficult. At the same time, the participation serves to extend one's understanding and provides opportunities for exploration. Volunteer activities also rely on a certain level of competence and effectiveness. Sustaining the efforts also requires that one learn to be sensitive to depleting one's attentional resources (i.e., increasing mental fatigue), although the chance to work in a natural setting can help to offset mental fatigue by drawing on the effortless kind of attention.

Information Sharing

Many of the topics discussed in this chapter involve dimensions of daily life that one takes for granted. People depend on information, build mental models and share their models with others; they like to feel competent and clear-headed and they care to make a difference. With all the experience that daily life provides one would think such actions would work smoothly and effectively. All too often, however, this is not the case.

Information sharing and collaboration turn out to be far more treacherous than any of us wants to admit. People are often more enthusiastic about giving information than receiving it. They like to talk, but are not as good at listening. They assume that what they say is what the listener heard. They readily consider their knowledge more pertinent than that of others. They are also more sensitive to their own feelings—especially negative ones—than to other people's feelings. All of us have been on both sides of these situations; giving too much information, yet being impatient with information that others try to provide. Wondering why the recipients of our good efforts can be so obtuse about understanding the issues. And also, frustrated that someone who provides information is making it so hard to follow the argument. These experiences too are a part of daily life.

There is considerable literature on the difficulties of such efforts. Efforts to communicate in interdisciplinary contexts are particularly challenging (Hodgson and Smith 2007; Marzano et al. 2006). There are also plenty of examples where good intentions to communicate fall short of the mark. Warburton et al. (2005) provide useful imagery from their interviews with environmental volunteers. The authors categorize the results in terms of four informational domains, including information that is too specialized, difficult to understand, and delivered by a coordinator (which reduced opportunities for exploration and self-pacing). As the authors indicate, "Information may be available to groups, but it is often too complex, too voluminous, or too technical, thus encouraging a schism between an increasingly professionalised scientific community and holders of local environmental knowledge" (Warburton et al. 2005, p. 28). From the perspective of the RPM one could describe the failures of the process in terms of participants' difficulties in model building, by undermining both understanding and efforts to explore on their own terms. This in turn reduces their sense of competence, and diminishes their ability to make a difference.

While the mechanisms for acquiring, storing, and retrieving information are common for the species, the mental models themselves are not. Model building is an individual activity; no plug-ins or flash drives can do it. Worse yet, it is a relatively slow process, generally requiring considerable time and experience. The experience, in turn, impacts the acquisition and use of subsequent information in ways that are subtle, yet pervasive. Another way to think of such accumulated experience is in terms of expertise—a degree of understanding that is necessary and often expected. It turns out however, that expertise is a mixed blessing. It is a key issue in considering information sharing.

Expertise

An "expert" traditionally refers to someone who has extensive skills and knowledge acquired through prolonged training and experience. In some areas there are specified standards for expert designation, including certification or licensing. Experts are assumed to be proficient and experienced with respect to their area of expertise.

If expertise accrues with experience in a particular field or area, then all of us, even children, have substantial expertise. Children may be expert readers; they are experts about where things are in their room, home, and school; they may even have comfort and ease of doing some things adults in the same household lack, such as computer skills or deciphering a "secret" code system.

Whether the context is professional expertise or the kinds everyone has, the extensive experience that leads to the expertise has some concomitant characteristics that are likely to be barriers to information exchange. In particular, one assumes that one's perception is of a "true reality," and, therefore, others must also see what is right there. Closely related is how readily one forgets what it was like before acquiring the expertise. Expertise leads to a different way of seeing—or perceiving, or recognizing, or organizing—the information.

De Groot's (1965) studies of expert chess players provide vivid demonstration of these differences. He asked novices, master players, and chess experts to view chessboard configurations for 5 seconds and then reconstruct them from memory. When shown random arrangements on the board, de Groot found that the groups were not particularly different in their performance. With meaningful arrangement of the chess pieces, however, the situation changed dramatically. Now, only the experts showed great facility. The studies showed that the differences were not a matter of memorization, but rather, expertise leads to differences in the ability to see patterns. Pattern recognition, however, is not usually a part of our awareness. We are not aware of patterns of letters forming the words we read, nor the notational system (i.e., punctuation) that guides our comprehension of paragraphs. Looking at text in a foreign language with a different alphabet, however, we quickly realize our deficit.

Exchanging information depends on these patterns. The patterns one sees in a "wetland" depend on experience. Recognition of a bog or fen or a particular kind of marsh is an acquired reality that calls on the seemingly instant awareness of combinations of factors. The patterns one sees in a remote sensing image or a computer model are also acquired realities that capture diverse information in highly efficient form. These realities, however, were not always there, and are not likely to be available to those who do not share the pertinent expertise.

To an extent far greater than one realizes we all live in very different worlds. What we see, hear, and remember about the same circumstances can differ substantially. The "facts" in a specific situation as portrayed by a long-term resident, a developer, an ecologist, and a hydrologist may draw on a common vocabulary, but in other respects have little in common. In many other instances even the vocabulary may be unshared as each party accuses the other of resorting to jargon—terminology that is a convenient short hand for one group and a smoke screen as viewed by those who do not share it. Such failures of exchanging information often happen with no ill intent, no desire to deceive or withhold, to mystify or impress. They are the consequences of the natural expression of the way each party perceives the situation, necessarily but invisibly colored by prior experiences, training, and knowledge.

Antidotes

Expertise is thus a mixed blessing—necessary, desirable and valued, but also a potential obstacle to communication. Information sharing is also necessary, desirable, and valued. Finding ways to reduce the obstacles inherent in expertise is essential.

RPM tells us that people are not at their best when they are confused or overwhelmed—a situation that happens often when trying to follow information where one knows the words, but has little sense of what is said. RPM tells us people care to make a difference, yet many information sharing situations seem more like information dumps than exchanges. RPM also says that people depend on understanding and exploration to build and use their mental models. If they cannot understand what is going on and the path to exploration is impeded, then model building is stymied for all parties to the communication.

Limited Capacity

A major difference between one who knows (i.e., an expert) and one who is learning expresses itself in differences in the quantity of information that each can handle. In both instances there are limitations in the capacity to handle additional information. However, in areas where one has greater familiarity, the existing knowledge base is available in a more organized fashion, leaving more capacity available for dealing with new information.

This discrepancy in how much capacity is available leads to one of the most detrimental mismatches in efforts to exchange information. All of us have been overwhelmed by too much information or too many choices and thus have first-hand experience with the distress this can cause. Yet it is difficult to recognize that information that seems straightforward to us may be far more mystifying to someone lacking the needed background. The temptations to add one more fascinating aspect, to include yet another line of support, or to show one more amazing ramification are difficult to resist. Yet, the cost of these seemingly small embellishments can be colossal in terms of the intended information exchange.

The solution to this problem involves holding back. If the consequence of too much information is likely to be that none of it is remembered, then providing far less information will yield better results. Less can be more.

Another part of the solution is in the way the information is structured. How often does one remember even three main points from a presentation? How rarely are presentations delivered mindful of such a goal? Structure is important in written material as well. Long lists have little chance of successful information transmittal. However, organizing key points and providing a few sub-themes can work quite effectively. In other words, attention to hierarchical organization of the issues speaks to the dilemma that limited capacity poses.

Structure can be communicated visually with headings and "white space." Orally, indications of structure are provided by signaling transitions, by orienting the listener to where one has been and where the story is going. Structure helps one see that there are a manageable number of parts to the whole, and to keep track of where one is along the story's path. These all are helpful for understanding.

Forms of Engagement

Finding ways to engage others is pivotal to information sharing. Ironically such efforts are frequently ignored especially in contexts where technical material is at issue. Exploration, an important dimension of building mental models, is closely related to interest. Enticing the reader or listener to get deeper into the information, to wonder, to seek the resolution to uncertainties that have been created—are all devices that can enhance exploration.

Starting with a question or a paradox can increase curiosity and the desire to find out. While such devices can be misused and overused (e.g., a long list of questions), their total absence in technical materials is an unfortunate omission. A question requires an action on the part of the recipient. It can signal an opportunity for exploring possibilities or indicate the intention for an exchange of information rather than a one-sided offering.

Stories and examples can also be effective ways to engage others. They offer concreteness which can help clarify specific issues that are pertinent for the situation at hand. Furthermore, since mental models are built from many instances the use of example can broaden understanding of the problem space and offer ways to explore alternative approaches.

Visual material is also a potentially effective way to engage. A picture *can* be worth a thousand words. However, visual material may also fail in its good intentions. It can be entertaining without contributing to the substantive intention. It can fall prey to the expertise dilemma by being overwhelming and thus confusing.

In addition to graphics or pictures, attention to the spatial arrangement of the information can also be important for engaging a reader. Consideration of the layout (e.g., white space on the page) is not only pertinent for public relations and selling; it also can help to make the material more readily understandable. Grayness of a page and density of the content make it less inviting and add to the communication barriers even if the material is well written.

The "Where They're At" Principle

It is quite frustrating to have information one provides be dismissed or tuned out. People can be quite finicky about receiving information. They are not eager for information they think they already know. They are not welcoming of information that is contradictory to what they hold dear. They resist what might appear as irrelevant, boring, or threatening.

These difficulties suggest the importance of relating information to "where they're at." By trying to connect information to the mental models and perceptions already in the heads of the intended recipients, the information has a greater chance to connect to what is pertinent; the effort thus turns familiarity to advantage, rather than permitting it to get in the way. Such efforts are useful not only for facilitating model building, but also speak to the meaningful action aspect of the RPM. Engaging with what the other person already knows and is concerned about is a sign of respect and provides a chance for participation.

In many situations it may be difficult to ascertain people's familiarity. For example, brochures and other written material are likely to reach diverse audiences. A public presentation may also provide limited opportunities for appreciating the diverse perspectives of the audience. Even in such situations, however, one can orient information so as to build from existing knowledge and concerns. For example, an information booklet can indicate that the news to be offered may be unwanted or that the information may seem like what the reader already knows but that there is a twist. One can ask the recipients to consider their own perceptions of the situation before proceeding to impart information. In that way, there might be greater recognition of discrepancies between what is offered and what might have been considered obvious. In other words, rather than assuming that readers or listeners are eager for whatever is offered, attention to their knowledge, worries, and circumstances can lead to more effective information exchange.

Summary

There are many reasons why information exchanges often fare badly. The issue of familiarity is central to effective information exchange, but it can also be a source of missteps and predicaments. One's own familiarity—one's expertise—can be particularly problematic as a barometer of information to offer. Expertise challenges our ability to judge the amount of information to offer and blinds us to the different way we see situations.

The recipients' familiarity with information can also get in the way of effective exchange. They may ignore information that is cast as "nothing new" and shun information feared to conflict with their existing mental models. While such reactions are understandable they are nonetheless roadblocks in efforts to exchanging information.

Fortunately, information sharing can also go well. To the extent that people can draw on what is familiar to them, or be helped to reduce the unfamiliarity of a new situation, the process has a greater chance of succeeding. Finding common ground, building on what the listener or reader is likely to find familiar and engaging, and concern to keep the amount of information manageable are all important elements in facilitating the process.

Some simple, yet far-reaching observations to keep in mind:

- Valuable information is not the exclusive property of those with expertise or status;
- Multiple perspectives can be enriching and lead to constructive outcomes;
- Information in the hand is not information in the mind; and
- Humans are sensitive to signs of making a difference (as well as to their absence).

Nearby Nature and Well-Being

The second major theme of the chapter focuses on wetlands as a part of the landscape. While the context is quite different from information sharing, many of the same issues apply here as well. A wetland, just like a chessboard, consists of many patterns. The patterns that different people see are the product of their experience. The ecologist will see different patterns than the artist. The local resident will see different patterns than the engineer. Not only do all these individuals trust what they see, but they depend on these perceptions for their actions and passions.

These actions and passions are vitally important. A successful project depends on them. At the same time, however, such actions and passions can also be the source of unexpected hostility and controversy. Ecological restoration projects of various kinds have been met by disagreements and antagonism (Marler et al. 2005; Ostergren et al. 2008). Well-intended actions by knowledgeable professional have received outraged responses and resentment from local citizens. In such confrontations professionals are likely to feel that their sound judgment is questioned by people lacking the necessary training. The public, in turn, feels that their treasured nearby environment is violated by outsiders who have no understanding for what it means to them. In the case of the "Chicago restoration controversy" (Gobster 2000), the emotional storm was sufficiently intense to lead to a moratorium on all restoration practices; some of the restrictions were sustained for a decade. What these unfortunate situations exemplify is not only the diversity of perspectives, but also the intensity of people's passion for the natural environment (Kaplan and Kaplan 1989).

It is important to acknowledge the differences in perspective. However, recognizing commonalities is also crucial; these may be particularly helpful in avoiding negativity and finding solutions. The three domains of the RPM point to some of these commonalities.

Meaningful Action

The professionals' frustrations are intensified by the sense that their knowledge is not respected. The public's anger is exacerbated by their sense that they are not listened to. Reasonableness is more likely when people think that they can make a difference. To bring out the best in all the parties involved thus requires a process that is respectful, genuinely participatory, and inclusive. Phalen's (2009) comparison of two ecological restoration projects highlights the vital role the participatory process can play. Many of the issues raised in the context of information sharing are central to the participatory process as well. Participation entails both informing and engaging. Not only does the public need to be "educated" about the issues, but the various professionals (e.g., decision makers, engineers, ecologists, designers) can also benefit from the information exchange.

The fact that people want to participate in the protection and preservation of wetlands is an important indication of meaningful action. Many thousands of acres of wetlands have benefitted from the involvement of volunteers. The benefits, however, extend beyond the wetlands, to those who participate in the activities and even beyond them. The volunteers' dedication improves a local resource as well as their community. Others in the community gain pride in the local surroundings and the enhanced sense of community can lead to further activities and initiatives. At the same time there are gains for those who volunteer. They expand their knowledge and skills. They find new interests and new friends. Even more important than these personal benefits, however, is the satisfaction gained from feeling that they can make a difference with respect to something that matters to them. A number of studies have found that items related to "helping the environment" are the most highly endorsed motivations for (or benefits from) participating in environmental stewardship activities (Bruyere and Rappe 2007; Grese et al. 2000; Ryan et al. 2001).

Being Effective

Recall the earlier discussion of the effortless kind of attention that is so essential to mental restoration. While many kinds of settings can serve to replenish the depleted attentional capacity, the preponderance of research, however, has focused on the role played by the natural environment in this context (Frumkin 2005). This is where wetlands play a particularly important role. For many people, whether or not they are personally involved in restoration efforts, these are settings that offer enormous fascination. The landscape itself, the wildlife and birds, the subtle variations in hues and textures all enrich the experience. Participation in the landscape comes from diverse ways of relating to it. It may involve physical activities (e.g., fishing, canoeing, or hiking); it can also entail observation, even from afar (e.g., birding or even just observing from a window). Whether in the setting or viewing it from a distance, the experience can foster mental clarity and restoration.

While the various professionals involved in wetland restoration and resources see the wetland differently, they too may experience the subtle benefits that are afforded by the natural resource. Watching a wetland take shape, monitoring its progression, and even knowing of its availability can all contribute to mental restoration.

Model Building

Research on environmental preference led to our realization of how pivotal understanding and exploration are to people's perception of the environment (Kaplan and Kaplan 1978, 1989). The research showed surprising agreement across many kinds of settings and variations in participants' experiences. Across these diverse contexts, study participants consistently show lower preferences for scenes that suggest their path will be obstructed or ones that have great homogeneity across the entire setting. By contrast, preferences are routinely high when scenes suggest that one might learn more as one can move deeper into the pictured space. Feeling reassured that that one would be able to find one's way also plays an important role.

We found from these studies that people's preferences are not whims or eccentric matters of taste. Rather preferences are an indication of people's assessment of their ability to negotiate the environment. What is this environment about? How can I make my way through it? Will I find my way there and back? What will I learn as I move through the space? Implicitly, these are the questions that we all ask ourselves all the time in many contexts. Consider a presentation as an environment, or a new computer program you are trying to learn, or getting ready for a trip to another city. Questions such as these reflect people's efforts to understand an environment (or situation or setting) as well as their assessment of expanding that understanding by exploring.

Wetlands are no different. Whether considering wetlands in general or a specific location, one's mental models guide the assessment. Without knowledge of wetlands, a person may look at the expanse and consider it lacking in variation and difficult to negotiate. With experience the same expanse can reveal a great deal of variation. If the experience is based on frequent exposure to a particular marsh, for instance, if it is included in one's view from home, then one's mental models permit one to predict likely events (e.g., migratory birds) or color changes. Similarly, the mental models of individuals who volunteer in restoration efforts are enriched by their greater understanding and exploration of the site.

Knowing one wetland extremely well is different from knowing a great deal about the hydrology of wetlands in general. In other words, the mental models of a local resident and of a hydrologist will be very different. Both are guided by their mental models, but their models can lead them to different concerns and considerations. It is hard to appreciate that these mental models, though very different, serve to guide behavior and are central to people's affections and trepidations. It is particularly difficult to appreciate these differences as they are not generally evident. In fact, they are most likely to emerge when there are disagreements.

Such disagreements signal how much the particular environment matters to people. As Ryan (2005) has shown, the strength of attachment may not be realized until there is a threat. For a local resident or recreationist a restoration project is about a specific place—perhaps an area where one goes birding or it may be part of the view from home. Changes to such familiar places can undermine one's comfort and reduce one's opportunities for personal restoration. For the various professionals involved in the project the site itself is unlikely to be the source of attachment; the important issues are that the job is done well.

Things that Matter

People are not likely to become upset about things that do not matter to them. When they are persistent or difficult or obstreperous it is often a clear indication that they care deeply. What people care about is intimately connected to their dependence on information. The context for the information constitutes an environment—the physical environment of the wetland, the maps, and diagrams that represent it, the conceptual space of exchanging ideas about it. All these play a role in how humans relate to each other and to the world around them.

People can be particularly problematic when:

- Environments hinder or block their **understanding**;
- Environments lack opportunities for **exploration**;
- Environments fail to foster experiences that are **restful and enjoyable**; and/or
- People feel their **participation** is not welcome.

These four topics (quoted from Kaplan et al. 1998, p. 149) are true for citizens of different ages, ethnicities, recreational preferences, and residential choices. They are also just as true for people with expertise in different domains that pertain to the problem at hand. The stronger one's passion about the wetland and its viability, the more it is worth achieving a desirable outcome. The stronger one's conviction about one's understanding of the issues, the more it is worth making sure that it is represented in the solution.

The more we all can recognize the human characteristics that we share, the easier it may be to foster reasonableness. Not only is information sharing critical to that outcome, but so are the wetlands themselves.

Acknowledgements The financial support provided by the United States Department of Agriculture, Forest Service Northern Research Station, is gratefully acknowledged. As always, Stephen Kaplan's contributions extend far beyond his helpful comments.

References

Bruyere B, Rappe S (2007) Identifying the motivation of environmental volunteers. J Environ Plan Manage 50:503–516

De Groot AD (1965) Thought and choice in chess. Mouton De Gruyter, The Hague, p 479

Frumkin H (2005) The health of places, the wealth of evidence. In: Barlett PF (ed) Urban place: reconnecting with the natural world. MIT Press, Cambridge, pp 253–270

Gobster PH (2000) Restoring nature: human actions, interactions, and reactions. In: Gobster PH, Hull RB (eds) Restoring nature: perspectives from the social sciences and humanities. Island Press, Washington DC, pp 1–19

Grese RE, Kaplan R, Ryan RL, Buxton J (2000) Psychological benefits of volunteering in stewardship programs. In: Gobster PH, Hull RB (eds) Restoring nature: perspectives from the social sciences and humanities. Island Press, Washington, pp 265–280

Hawken P (2007) Blessed unrest: how the largest movement in the world came into being and why no one saw it coming. Viking, New York, p 352

Hodgson SM, Smith JWN (2007) Building research agenda on water policy: an exploration of water framework directive as an interdisciplinary problem. Interdiscip Sci Rev 32:187–202

Kaplan S (1973) Cognitive maps in perception and thought. In: Downs RM, Stea D (eds) Image and environment. Aldine Transaction, Chicago, pp 63–78

Kaplan S (1995) The restorative benefits of nature: toward an integrative framework. J Environ Psychol 15:169–182

Kaplan S (2001) Meditation, restoration, and the management of mental fatigue. Environ Behav 33:480–506

Kaplan S, Kaplan R (1978) Humanscape: environments for people. Duxbury, Belmont (Republished by Ulrich's Books, Ann Arbor, 1982)

Kaplan R, Kaplan S (1989) The experience of nature: a psychological perspective. Cambridge University Press, New York (Republished by Ulrich's Books, Ann Arbor, 1995)

Kaplan R, Kaplan S (2005) Preference, restoration, and meaningful action in the context of nearby nature. In: Barlett PF (ed) Urban place: reconnecting with the natural world. MIT Press, Cambridge, pp 271–298

Kaplan S, Kaplan R (2009) Creating a larger role for environmental psychology: the reasonable person model as an integrative framework. J Environ Psychol 29:329–339

Kaplan R, Kaplan S, Ryan RL (1998) With people in mind: design and management of everyday nature. Island Press, Washington DC, p 239

Marler MJ, Supplee K, Wessner M, Marks G (2005) Changing attitudes about grassland conservation in Missoula, Montana—"Weed capital of the west." Ecol Restor 23:29–34

Marzano M, Carss DN, Bell S (2006) Working to make interdisciplinarity work: investing in communication and interpersonal relationships. J Agric Econ 57:185–197

Ostergren DM, Abrams JB, Lowe KA (2008) Fire in the forest: public perceptions of ecological restoration in north-central Arizona. Ecol Restor 26:51–60

Phalen KB (2009) An invitation for public participation in ecological restoration: the reasonable person model. Ecol Restor 27:178–186

Ryan RL (2005) Exploring the effects of environmental experience on attachment to urban natural areas. Environ Behav 37:3–42

Ryan RL, Kaplan R, Grese RE (2001) Predicting volunteer commitment in environmental stewardship programmes. J Environ Plan Manage 44:629–648

Simon HA (1971) Designing organizations for an information-rich world. In: Greenberger M (ed) Computers, communication, and the public interest. The Johns Hopkins Press, Baltimore, pp 37–72

Warburton JS, Marshall K, Warburton, Gooch M (2005) Information-related constraints on the effectiveness of environmental volunteers: a case study. Aust J Volunt 10:24–31

Chapter 10
Wetlands Regulation: The Case of Mitigation Under Section 404 of the Clean Water Act

Morgan Robertson and Palmer Hough

Abstract The requirement to mitigate impacts to wetlands and streams is a frequently-misunderstood policy with a long and complicated history. We narrate the history of mitigation since the inception of the Clean Water Act Section 404 permit program in 1972, through struggles between the United States Environmental Protection Agency and the United States Army Corps of Engineers (ACOE), through the emerging importance of wetland conservation on the American political landscape, and through the rise of market-based approaches to environmental policy. Mitigation, as it is understood today, was not initially foreseen as a component of the Section 404 permitting program, but was adapted from 1978 regulations issued by the Council on Environmental Quality as a way of replacing the functions of filled wetlands where permit denials were unlikely. The Environmental Protection Agency (USEPA) and the ACOE agreed in 1990 to define mitigation as the three steps of avoidance, minimization, and compensation, principles which must be applied to permit decisions in the form of the environmental criteria in USEPA's 404(b)(1) Guidelines. Through the 1980s and 1990s, the compensation component of mitigation has become nearly the sole focus of mitigation policy development, and has been the subject of numerous guidances and memoranda since 1990. Avoidance and minimization have received far less policy attention, and this lack of policy development may represent a missed opportunity to implement effective wetland conservation.

The regulatory architecture that has been built around wetland protection in the United States is unmatched in its complexity. From the apparently simple requirement in the Federal Clean Water Act (CWA) that a permit be required for the dispos-

---

Parts of this chapter appeared in *Wetlands Ecology and Management* 17:15–33, 2009.

M. Robertson (✉)
Department of Geography, University of Kentucky, 1457 Patterson Office Tower, Lexington, KY 40506, USA
e-mail: mmrobertson@uky.edu

B. A. LePage (ed.), *Wetlands,*
DOI 10.1007/978-94-007-0551-7_10, © Springer Science+Business Media B.V. 2011

al of dredge or fill material into a *"water of the United States"*[1] has grown a federal regulatory program of enormous scope, scale, and complexity. Additionally, various state and local governments have adopted parallel wetland protection programs, and other federal-level agencies such as the United States Department of Agriculture (USDA) and the National Oceanic and Atmospheric Administration (NOAA) have developed their own wetlands protection programs. This chapter cannot hope to treat the topic of wetland regulation with anything like a comprehensive view. Even the CWA program (described in Section 404 of the Act) can be divided into separate and equally controversial areas that have each generated litigation, scientific debate, and policy schisms. What is a "water of the United States"? What counts as "disposal"? What constitutes "fill"? What environmental criteria should be evaluated when a permit is issued? Many authors have attempted to survey parts of these questions, and while none has comprehensively described the evolution of wetland regulation, reading them together one can come to a broad understanding (see especially Houck 1989; National Research Council 1995; Downing et al. 2003). Here, we attempt to address just one aspect of wetland regulation, but it is one that touches on nearly all other elements of the CWA and thus can serve as a window on several areas of regulatory controversy and policy development. The general policy question that we focus on here is this: *Given that adverse impacts to wetlands will continue to be permitted, how can the permit program itself limit or repair the damage?* Leaving aside the complicated questions of jurisdiction, types of impact, and the relative worth (economic and societal) of wetlands to society, how does a regulatory author-

---

[1] As defined in 40 CFR 230.3(s) The term waters of the United States means:

a) All waters which are currently used, or were used in the past, or may be susceptible to use in interstate or foreign commerce, including all waters which are subject to the ebb and flow of the tide;

b) All interstate waters including interstate wetlands;

c) All other waters such as intrastate lakes, rivers, streams (including intermittent streams), mudflats, sandflats, wetlands, sloughs, prairie potholes, wet meadows, playa lakes, or natural ponds, the use, degradation or destruction of which could affect interstate or foreign commerce including any such waters:

   (i)   Which are or could be used by interstate or foreign travelers for recreational or other purposes; or

   (ii) From which fish or shellfish are or could be taken and sold in interstate or foreign commerce; or

   (iii) Which are used or could be used for industrial purposes by industries in interstate commerce;

d) All impoundments of waters otherwise defined as waters of the United States under this definition;

e) Tributaries of waters identified in paragraphs (s)(a) through (d) of this section;

f) The territorial sea;

g) Wetlands adjacent to waters (other than waters that are themselves wetlands) identified in paragraphs (s)(a) through (f) of this section; waste treatment systems, including treatment ponds or lagoons designed to meet the requirements of CWA (other than cooling ponds as defined in 40 CFR 423.11(m) which also meet the criteria of this definition) are not waters of the United States.

ity make society and the environment *whole* again, when temporary and permanent impacts to wetlands are allowed to occur? In the United States regulatory system, this is known as *mitigation*.

There are few words in the lexicon of wetlands regulation in the United States more freighted with baggage than mitigation, and fewer still are so commonly misused and misunderstood. Section 404 of the 1972 Federal Water Pollution Control Act (FWPCA) established a permit program to regulate the discharge of dredged or fill material into "waters of the United States," which includes jurisdictional wetlands. Congress divided the responsibilities for Section 404 between the United States Army Corps of Engineers (ACOE) and the United States Environmental Protection Agency (USEPA). The ACOE was tasked with the day-to-day administration of the permit program, issuing permits for regulated activities in the nation's jurisdictional waters. The USEPA developed the environmental criteria used by the ACOE to make its permit decisions and shares enforcement authority with the ACOE for Section 404. In this partnership, the USEPA and the ACOE developed national Section 404 mitigation requirements and policy in close collaboration; however, forging a common vision and interpretation has not been easy. The following history of wetlands mitigation under Section 404 is largely the story of these two federal partners attempting to bring together their divergent missions and divergent constituencies to serve the common and most basic need to protect the nation's wetlands.

The Early Years of Wetland Mitigation

In retrospect, it has been only 15 years since the concept of wetland mitigation was first proposed as a permit stipulation. The initial concept (acquisition and preservation of undeveloped wetlands in exchange for permits to develop other wetlands) evolved significantly before it was codified as a written document. In that relatively brief time period, we have generally succeeded in establishing the legitimacy of the concept. (LaRoe 1986, p. 9)

The FWPCA was originally passed in 1948, primarily as a bill to fund the construction of municipal water treatment works and encourage other voluntary measures to promote hygiene. The 1972 amendments, however, transformed the Act from a funding vehicle to a regulatory mechanism. The most significant transformation for our purposes was the establishment of a permit program in Section 404 that regulates the discharge of dredged or fill material into "waters of the United States". The concept of mitigation for permitted impacts was mentioned in ACOE regulations as early as 1973 (United States Army Corps of Engineers 1973). However, given the consuming nature of fundamental questions concerning the extent of CWA jurisdiction (*cf.* Wood 2004), the application of mitigation requirements was not immediately a central issue after the passage of the 1972 CWA. Furthermore, mitigation may not have initially been seen as a priority if it was assumed that permits for work, which truly damaged wetlands would either be denied by the ACOE or "vetoed" by the USEPA under its Section 404(c) powers. While this position seems unlikely today, mitigation for permitted wetland impacts was not specifically men-

tioned when the USEPA developed the environmental criteria for the issuance of
ACOE 404 permits in 1975 (United States Environmental Protection Agency 1975).
However, the considerable changes made to the permit program in the 1977 amendments made the question of mitigation unavoidable. The affirmation of the use of
General Permits by the ACOE was an acknowledgement that Congress intended to
allow large numbers of "minor" permitted impacts which damaged wetlands. General Permits made the permitting process manageable for the ACOE staff and less
time-consuming for some applicants, but also begged the question of how so many
permits could be issued while still achieving the statutory goals of the CWA, that
is, assuring the biological, chemical, and physical integrity of the nation's waters.
As questions of jurisdiction and workload began to recede, attention turned to the
process of mitigating the effects of massive numbers of permitted impacts.

Kruczynski (1990) notes that some mitigation was performed in association with
permits in the 1970s, but was not sanguine about it. "[A]s early as 1975 agencies
would compromise their positions on a permit application as long as there was, at
least on paper, no net loss of wetlands. Federal agencies recommended compensatory replacement mitigation, in part, due to the USEPA's hesitancy to use its Section 404(c) authority [to veto the issuance of ACOE permits]" (Kruczynski 1990,
p. 551). Thus, the practice of mitigation grew as a consequence of the agencies'
minimal use of their CWA powers: the ACOE's unwillingness to deny permits that
entailed significant environmental damage, and the USEPA's unwillingness to veto
such permits.

At this time, and in fact throughout much of the 1980s, it was the United States
Fish and Wildlife Service (USFWS) and the National Marine Fisheries Service
(NMFS) in coastal areas that was more empowered to request that mitigation measures be attached to permits. This was an exercise of their authority under the Fish
and Wildlife Coordination Act of 1939, and it was used extensively (LaRoe 1986).
A NMFS survey in 1981 showed that the NMFS commented on 22% of the Section 404 permit applications that it reviewed, and that its mitigation recommendations were incorporated 98% of the time (Hall 1988). Often these mitigation measures took a form that is now termed "compensation," using the new technology of
marsh creation, which had developed from the successes of the ACOE's Dredged
Material Research Program (Webb et al. 1986). However, Kruczynski notes that the
early successes of these projects in tidal areas were used to justify issuing permits
for impacts to wetland ecosystems that were not as easily restored (such as bogs,
fens, and bottomland hardwood swamps), and with a less well-documented history
of technical experimentation (Kruczynski 1990, p. 552).

The Mitigation Sequence

In CWA parlance, "the mitigation sequence" consists of the procedural steps where
decisions on the level of impact and of appropriate mitigation are made. Daily
practice and federal guidance tell us that mitigation consists of three steps: impact
avoidance, impact *minimization*, and impact *compensation*, to be achieved in that

order. Terms like "mitigation" and "minimization," and the concept of an alternatives analysis that prioritizes avoidance, had appeared in policy debates at the state and federal level throughout the 1970s (LaRoe 1978). However, they were not all brought together in a structured way until the Council on Environmental Quality (CEQ) clarified the National Environmental Policy Act (NEPA) regulations in 1978. Section 1508.20 of these regulations defines "Mitigation":

> Mitigation includes: (a) Avoiding the impact altogether by not taking a certain action or parts of an action. (b) Minimizing impacts by limiting the degree or magnitude of the action and its implementation. (c) Rectifying the impact by repairing, rehabilitating, or restoring the affected environment. (d) Reducing or eliminating the impact over time by preservation and maintenance operations during the life of the action. (e) Compensating for the impact by replacing or providing substitute resources or environments.

Soon afterwards, the USEPA issued a revision of the 1975 environmental criteria. The new regulations, known as the Section 404(b)(1) Guidelines, were issued in a final form on December 24, 1980 (United States Environmental Protection Agency 1980). Despite the sobriquet "Guidelines," they are an enforceable regulation. The preamble to the proposed Guidelines (44 FR 54223) affirms that mitigating impacts is required (not optional) by pointing to two existing statutes: the CWA's statutory requirement that the 404 permit program be "based on criteria comparable to" the Section 403 (mandatory) ocean discharge criteria, and NEPA's concept of an alternatives analysis, which must apply by the CEQ's 1978 rule, even to small federal projects that do not require a full Environmental Assessment under NEPA.

The Guidelines define mitigation by constructing a series of prohibitions and rebuttable presumptions that, taken together, mandates a *sequence* of mitigation steps that must be followed when issuing and conditioning a permit. An "alternatives test" designed to identify the least environmentally damaging practicable alternative (LEDPA) must come *before* efforts that address unavoidable impacts. This mitigation sequence is contained in the requirements listed in Section 230.10 of the Guidelines.

To achieve mitigation through *avoidance*, the ACOE must consider the available alternatives to the proposed action. If the proposed action involves non-water dependent impacts (e.g., building homes, shopping malls, highways, etc.), then Section 230.10(a)(3) makes the alternatives analysis more rigorous by establishing two rebuttable presumptions. First, it is presumed that alternatives that do not impact aquatic resources are available and feasible. Second, it is presumed that such alternatives are environmentally preferable. Both presumptions may be rebutted by the applicant. Having established the LEDPA in Sections 230.10(a)–(c), the permittee then moves to mitigation by *minimization* and *compensation* in 230.10(d).

It is not clear when the notion of three distinct steps first arose, and how they are distinct from the five-part definition articulated by CEQ in 1978. As early 1982, Race and Christie had stated:

> Many commentators tend to apply the term mitigation to three categories that can be described as (1) planning to prevent damage to the environment, (2) design and execution of projects to minimize adverse impacts, and (3) restoration or compensation for unavoidable damage to the environment. New definitions are continually generated and manipu-

lated to suit the purposes of a given author, developer, or regulator. However, all of the definitions expand upon the traditional definition of mitigation, which focuses on *post facto* actions taken to restore or compensate for the unavoidable impacts of an activity on wetlands. The evolution of the definition is understandable; it is disagreeable to many people to assume damage without some attempt at protection first. (Race and Christie 1982, p. 318)[2]

Another sequential approach to the steps of mitigation was contained in the 1981 USFWS mitigation policy (United States Fish and Wildlife Service 1981). The USFWS policy, used in formulating mitigation recommendations for the ACOE permit applications, stated that "These means and measures are presented in the general order and priority in which they should be recommended by Service personnel..." (United States Fish and Wildlife Service 1981, p. 7,660), and retained the five CEQ mitigation categories of Avoidance, Minimization, Rectification, Reduction, and Compensation. However, the application of these categories as a sequence was couched within four categories of impact. Resource Category 1 is afforded the highest level of protection and only insignificant impacts should be permitted. The mitigation sequence only applies to Resource Categories 2 through 4, while the requirement that any compensation be of the same habitat type (the "in-kind" preference) is progressively loosened from Resource Category 2 to 4.

The ACOE and USEPA signed a Memorandum of Agreement on mitigation in February 1990 (United States Army Corps of Engineers and United States Environmental Protection Agency 1990). It interprets key provisions in the Section 404(b) (1) Guidelines to establish the policies and procedures to be followed in determining what mitigation is necessary for compliance with Section 404, as well as the three-step sequence itself. For reasons clarified in the next section, it is impossible to overstate the importance of the 1990 MOA in reframing the entire debate over mitigation, and in giving us the current meaning of nearly all of the concepts that characterize mitigation. In the MOA, the USEPA and the ACOE agree that mitigation for standard permits proceeds in a sequence: avoid, minimize, and compensate. It also incorporates the preferences for "in-kind" compensation and for compensation to occur on or adjacent to the impact site (i.e., "on-site" compensation) that had first been articulated by the USFWS in 1981.

The MOA also put boundaries on the concept of mitigation: its application has limits and it cannot cure all ills. It stipulates that some projects have impacts that are "so significant that even if alternatives are not available, the discharge may not be permitted regardless of the compensatory mitigation proposed" (United States Army Corps of Engineers and United States Environmental Protection Agency 1990, p. 4). Permit denials are rare enough (only 0.48% of all permit applications were denied in 2007 and 2008)[3], but the language is strong and may acquire more practical meaning in the future.

---

[2] "This article helped memorialize the description as a 3-step process. For a time, everyone who wrote on the subject after this article used the Race and Christie description or a slight variation of it." EPA Region 1 staff member, personal communication 6/8/07.

[3] Personal communication between William James and Brian Frazer 6-12-2009.

Avoidance: Attleboro Mall and Plantation Landing

Although the ACOE had initially resisted using the USEPA's guidelines when issuing permits, by the mid 1980's the ACOE and USEPA had come to an agreement that the USEPA's mitigation regulations were binding on the process of issuing ACOE permits. But what do "avoid, minimize, and compensate" mean on the ground? The meaning of "avoidance" was determined in two late-1980s cases: Attleboro Mall (Massachusetts) and Plantation Landing (Louisiana). The Attleboro Mall case concerned a permit application to construct a shopping mall in a wetland known as Sweeden's Swamp. This case looms large in the history of the Section 404 program for numerous reasons. It is one of only 12 cases in the history of the Section 404 program in which the USEPA has used its 404(c) "veto" authority over a ACOE permit decision.[4] The USEPA's Final Determination for this veto action (United States Environmental Protection Agency 1986) is most often remembered for its aggressive stance on impact avoidance through application of the "market entry" principle.[5] It is also critical for its affirmation of the sequential relationship between the Guidelines' requirement to avoid, minimize, and compensate for impacts. In making their application to the ACOE the permit applicant argued (and ACOE Headquarters agreed) that, when the compensation proposal was considered *simultaneously with* its proposed impact, the Sweeden's Swamp site was the least damaging alternative. In its veto of the permit decision, the USEPA ruled that the LEDPA must be determined *before* compensatory mitigation measures are considered, and that compensation measures are only encouraged "when there are no practicable alternatives other than filling in a wetland for a particular project and the project does not cause significant degradation to aquatic resources" (United States Environmental Protection Agency 1986, p. 2) This clearly articulated the sequence, and after the USEPA's veto action in this case was upheld in 1988 in *Bersani v. USEPA* [674 F. Supp. 405], the ACOE and USEPA put an end to the controversial practice of "buying down the LEDPA" with compensatory mitigation by including the following provision in the 1990 Mitigation MOA:

> Compensatory mitigation may not be used as a method to reduce environmental impacts in the evaluation of the least environmentally damaging practicable alternatives for the purposes of requirements under Section 230.10(a). (United States Army Corps of Engineers and United States Environmental Protection Agency 1990, p. 3)

Of equal import is the landmark guidance produced by the ACOE in response to the USEPA's challenge or "elevation" under the provisions of Section 404(q) of the ACOE's New Orleans District permit decision in the Plantation Landing case. The 14 pages of General Kelly's "Plantation Landing" memo to the field (Kelly 1989) provide essential guidance on the determination of the project's "basic purpose".

---

[4] http://www.epa.gov/owow/wetlands/regs/404c.html.

[5] In applying for a permit to construct a shopping mall in a wetland known as Sweeden's Swamp, the Final Determination stated that the applicant must consider alternatives to the wetland fill that were available at the time the permit applicant entered the market for the site, rather than at the time the applicant applied for a permit. And since a less environmentally damaging nearby site had in fact been available at that time, the permit for the Sweeden's Swamp site must be denied.

The Plantation Landing memo makes it clear that the ACOE must determine and evaluate the "basic purpose" itself, without the influence of the applicant. This is crucial because it prevents the project applicant from, for example, defining the "basic project purpose" as "to build a luxury golf-course development on this tract of land." Such a purpose would severely restrict the range of practicable alternatives. The ACOE might instead find that the "basic project purpose" is "to provide housing," and note that non-aquatic sites are available to fulfill this purpose.

In short, while the Attleboro Mall veto affirms the mitigation sequence, the Plantation Landing guidance defines all of the contextual information on purpose and alternatives needed to apply the mitigation sequence in specific situations. These were landmark moments that buried an important hatchet between the two agencies. When the 1990 "Mitigation MOA" came out the following year, it summarized and affirmed much of the substance and interpretation of these two cases.

Minimization in the Guidelines and in Practice

As previously noted, the Guidelines' *minimization* requirement is found in Section 230.10(d), which states that: "… no discharge of dredged or fill material shall be permitted unless appropriate and practicable steps have been taken which will minimize potential adverse impacts of the discharge on the aquatic ecosystem."[6] A variety of minimization measures are described in Section 230.75 (Subpart H) of the Guidelines including: changing the location of the discharge, changing the material to be discharged, controlling the material after discharge, changing the method of dispersion, changing the technology used, and changing the affects on plants, animals, and human uses.[7] Written prior to the development of low-impact design and the "green building" movement, many of these measures narrowly address the specific environmental impacts associated with the disposal of river and harbor dredge spoil. Only very general language is provided regarding minimization for activities such as residential, commercial, and industrial development activities that currently generate the vast majority of permit applications. This gap between the rather outdated language in the Guidelines and currently-feasible measures has created uncertainty regarding what actions can be required as "appropriate and practicable" minimization under the Guidelines.

Compensation

Compensatory mitigation is so central to discussions of mitigation that "compensation" is often mistakenly held to be synonymous with "mitigation," even among the most experienced observers of the program. It has been described as "the most

---

[6] 40 C.F.R. § 230.10(d) (2006).

[7] 40 C.F.R. § 230.70-77 (2006).

seductive concept in the field of wetlands protection" (Houck 1989, p. 836) because of the temptation to resolve tough permit decisions by seeking more aggressive compensation packages from permit applicants, rather than by fully exploring avoidance and minimization (Ciupek 1986; Yocom et al. 1989; Kruczynski 1990; Race and Fonseca 1996). These concerns are only exacerbated by the uneven track record of compensation site establishment and doubts regarding the ability of compensatory mitigation to actually offset permitted losses (see a compilation of studies in the National Research Council 2001, p. 190). Nevertheless, the ACOE permits allow impacts to approximately 22,000 acres of wetlands and other aquatic resources each year. To offset these annual losses, permit recipients are required to provide between 40,000 and 60,000 acres of compensatory mitigation.[8] A recent report estimates that providing this compensation costs approximately $ 2.95 billion annually (Environmental Law Institute 2007).

Methods

The compensatory mitigation required by the resource agencies generally fits within four methods: the *establishment* of a new aquatic site; the *restoration* of a previously-existing aquatic site; the *enhancement* of an existing aquatic site's functions, or the *preservation* of an existing aquatic site (usually through acquisition). Annually, over 65% of compensation takes the form of restoration and enhancement (Wilkinson and Thomson 2006). The federal resource agencies have a long-standing preference for the use of restoration over the other methods of compensation because it has the greatest potential for replacing both lost aquatic resource functions and area (United States Army Corps of Engineers and United States Environmental Protection Agency 1990; United States Army Corps of Engineers et al. 1995; United States Army Corps of Engineers and United States Environmental Protection Agency 2002). The use of establishment has decreased over the last fifteen years due to concerns over a high project failure rate and the loss of significant upland habitat. There are also concerns with the use of enhancement, which can offer functional improvements, but does not replace lost acreage. Finally, preservation has often been viewed skeptically, because preservation replaces neither lost functions nor lost acreage, and thus does not contribute to meeting the "no net loss of wetlands" goal (see below).

Mechanisms

There are three mechanisms for providing compensatory mitigation under the Section 404 program: permittee-responsible mitigation, mitigation banks, and in-lieu

---

[8] Personal communication between Russell Kaiser and Palmer Hough 4-19-07.

fee mitigation. *Permittee-responsible mitigation* is the most traditional form of compensation and still represents the majority of the compensation acreage provided each year (Wilkinson and Thomson 2006). It involves compensation undertaken by a permittee (or a contractor hired by the permittee) for impacts resulting from a specific project. As its name suggests, the responsibility for completing the work and ensuring success remains with the permittee.

A *mitigation bank* is a wetland or stream compensation area that is set aside to compensate for multiple development activities. The amount of compensation a bank can offer is determined by quantifying the aquatic resources restored or created in terms of "credits." Permittees, upon approval by regulatory agencies, can purchase these credits to meet their compensatory mitigation requirements. The mitigation banker is ultimately responsible for the success of the compensation project.

The concept of creating large blocks of compensation in advance of impact was first mooted in the early 1970s (Gosselink et al. 1974; LaRoe 1978). The USFWS was the agency which proposed this mechanism first; throughout the 1980s, the USFWS had a more sophisticated approach to mitigation than either the USEPA or the ACOE (LaRoe 1986). While the USEPA and the ACOE battled over the jurisdictional extent of the permit program, the USFWS was developing mitigation banking guidance (United States Fish and Wildlife Service 1983) and wrestling with compensation site performance standards and credit determination using assessment methods such as the Habitat Evaluation Protocol. What the USFWS lacked was a mechanism to force the ACOE to use the USFWS mitigation recommendations (Brown 1989). For many years, for example, some ACOE districts considered offsite mitigation to be impracticable by definition, and so the ACOE often refused to require any USFWS compensation recommendations where on-site compensation was not possible (Soileau 1984, p. 2). This left the USFWS strongly-motivated to develop a practicable and efficient off-site compensation method, which led directly to the birth of wetland banking at the Lafayette Field Office in 1981, and to the USFWS 1983 banking guidance.

Under the USFWS guidance, the first banks were created in the early 1980's as non-commercial ventures by state departments of transportation, energy companies, and other large-scale permit applicants to satisfy their own projected compensation needs. The USEPA and the ACOE began to exert their authority over wetland banking by the end of the 1980's, and to push for the adoption of entrepreneurial wetland banking. The first fully entrepreneurial banking venture was permitted in December 1992. The series of reports on mitigation banking between 1992 and 1995 by the ACOE Institute for Water Resources (e.g., Brumbaugh and Reppert 1994; Shabman et al. 1994) gave further agency sanction to the practice. With the issuance of the 1995 Federal Banking Guidance (United States Army Corps of Engineers et al. 1995), third-party producers sensed that conditions were stable for profit to be made more reliably, and the number of entrepreneurial banks expanded dramatically (Robertson 2006). A 2005 inventory estimated that there are approximately 363 active mitigation banks, 75 sold-out banks, and an additional 169 proposed banks under review (Wilkinson and Thompson 2006). Seventy-eight percent of these banks are for-profit entrepreneurial ventures.

While the ecological performance of these banks is widely supposed to be higher than that of other forms of compensation, this is largely a matter of anecdote and there is little comparative data to support this claim (but see objective evaluations in Spieles 2005; Mack and Micacchion 2006; Robertson 2006, 2007; Ruhl and Salzman 2006; Reiss et al. 2007). However, it is indisputable that wetland banking has resolved many intractable problems associated with permittee-responsible mitigation. Most significantly, the consolidation of many compensation activities into one large site has made it possible for the ACOE regulators to easily monitor and evaluate compensation site compliance.

In-lieu fee mitigation occurs when a permittee provides funds to an in-lieu fee sponsor, generally a public agency or non-profit organization, to satisfy a compensation obligation. The in-lieu fee sponsor pools these funds and eventually uses them to construct compensation projects. As with mitigation banks, the in-lieu fee sponsor is responsible for the success of these compensation projects. The use of in-lieu-fee compensation expanded through the 1990s, but assumed a variety of different forms, with some ACOE districts requiring in-lieu fee providers to establish detailed agreements resembling mitigation banking agreements, while other districts approved the ad-hoc use of fees with no formal agreement in place concerning how the money was to be spent or requiring the future compensation sites to be identified or secured. Often the money was never spent, or it was raided by state governments in deficit, or the amount paid was later found to be inadequate to the development of appropriate compensation sites (Gardner 2000). The number of in-lieu fee programs dropped from a high in 2001 of 87 programs to a total of 46 programs in 2005 (Wilkinson and Thompson 2006), although this was due in part to the re-categorization of some in-lieu fee programs as banks.

While over half of compensatory mitigation completed each year continues to be permittee-responsible compensation, in recent years, the use of mitigation banks has rapidly expanded and these banks currently provide over one-third of the annual compensation acreage with in-lieu fee compensation providing an additional 8% (Wilkinson and Thompson 2006).

No Net Loss

The popular slogan "no net loss of wetlands" is widely known, but has no existence in federal regulation. As the Section 404 provisions were being formulated in the 1970s, no one knew the nature or magnitude of wetlands losses—only that they were large, continuing, and significant. The first National Wetlands Inventory (NWI) report was published in 1983, indicating that between the 1950s and 1970s, the continental US had lost an average of 439,000 acres per year (Frayer et al. 1983). This number clearly staggered some observers, and in 1986 USEPA Administrator Lee Thomas called on the Conservation Foundation to convene a National Wetlands Policy Forum (NWPF), to provide a multi-stakeholder, comprehensive set of recommendations to address the newly-quantified crisis of wetlands loss. Meeting in 1987, the Forum was led by former New Jersey Governor Thomas Kean, whose state had been the first to require "no net loss" of wetlands in 1985 (Kantor and Charette 1988). The NWPF resulted in a slender volume (Conservation Foundation 1988) recommending the adoption of a national policy of "no net loss" and long term net gain of wetlands. On June 8, 1989 when President Bush advocated the achievement of no net loss in the highest-profile speech on wetlands to date by an American President:

> …generations to follow will say of us 40 years from now… that sometime around 1989 things began to change and that we began to hold on to our parks and refuges and that we protected our species and that in that year the seeds of a new policy about our valuable wetlands were sown, a policy summed up in three simple words: "No net loss." (United States Government Printing Office 1989, p. 694)

Although it was widely discussed, no net loss of wetland's (NNL) only appearance in a successful bill was in the 1990 Water Resources Development Act, which made NNL part of the official mission of the ACOE—but specified no other requirement that it be achieved. The popularity of NNL rhetoric had the effect of highlighting compensation within the Section 404 permit program. NNL provided the key notion of a "net" accounting of wetlands loss, which directly focuses policy on the dominance of compensation. When environmentalists cheered the NNL policy, they acquiesced to the notion that wetland protection was not merely to be achieved through the denial of permits, or even the avoidance and minimization of impacts, but rather through allowing impacts and requiring compensation.

Recent Developments in Mitigation

The continued relevance of mitigation has been reflected in every recent effort to evaluate the Section 404 Program. The National Research Council's (2001) study of compensation practices and outcomes is the most comprehensive evaluation to date regarding compensatory mitigation. The report's primary conclusion is a sobering one: despite progress in the last 20 years, "the goal of no net loss of wetlands is not being met for wetland functions by the mitigation program" (National Research Council 2001, p. 2). The report provides a comprehensive inventory of the shortfalls of compensation and identifies a suite of technical, programmatic, and policy recommendations for the Federal agencies, States, and other parties involved in compensation.

In April 2008, in response to a 2003 Congressional directive, the USEPA and the ACOE issued a rule (United States Army Corps of Engineers and United States Environmental Protection Agency 2008) designed to improve compensation by proposing equivalent standards that apply to all forms of compensation. This rule, for the first time, provides enforceable standards and practices concerning compensation. It attempts to respond to the recommendations in the NRC (2001) report by requiring clear performance standards, administrative procedures, and the use of available wetland scientific knowledge. The rule's standards for all compensatory mitigation are similar to the provisions that have been in place for mitigation banks since the 1995 banking guidance, and include: the use of real estate instruments to protect the compensation site; the funding of financial assurances for near- and long-term site stewardship; implementation of monitoring and contingency planning; and the clear identification of parties responsible for project tasks. Though not without implementation flaws, mitigation banking was seen as a model to guide the reform of other compensation mechanisms because banking is the only compensation mechanism that is "performance-based". All other types of compensation involve impacts occurring before compensation sites have achieved any performance standards or satisfied any administrative criteria.

Wetland Mitigation Outside the United States

The concept of requiring mitigation for permitted impacts to wetlands has spread around the world as governments and international bodies such as the Ramsar Secretariat seek to preserve the functions and area of their wetland resources in the context of a national permitting program. The 1971 Ramsar Treaty itself (in Sect. 4.2), suggests that signatories "should" compensate for reducing the extent of wetlands listed as internationally significant under the treaty. This concept was elaborated by the signatories at conferences in 1990 and 1996 (Gardner 2003). However, the provisions of the Ramsar Treaty are not expressed in mandatory language, but rather are often preceded by "should" or "may". At sites listed as internationally significant by the Ramsar Treaty signatories, the concept of "no net loss" has been applied in some cases. This allows compensatory mitigation to be used where impacts to wetlands are unavoidable in, for example, the Mai Po Inner Deep Bay site in Hong Kong.

Gardner's (2003) summary of international wetland conservation regulations and practices also makes mention of international migratory bird treaties such as the Bonn Convention, the North American Waterfowl Management Plan, and the African-Eurasian Waterbirds Agreement, which have in some cases been used to achieve mitigation for wetland losses, but these likewise do not have the force of a regulatory program and mitigation is not a legal requirement by which signatories are bound.

National-level wetland mitigation policy can be found in many countries, however. Both Australia and New Zealand have adopted policies similar to the United States. New Zealand's policy of "avoid, remedy, mitigate" mirrors the US mitigation sequence of "avoid, minimize, compensate", and has been applied to coastal marine areas by the Resources Management Act of 1991, although the measures are not clearly sequential in the Act. Various Australian states such as Victoria and New South Wales (NSW) have adopted mitigation practices in strict regulations protecting native vegetation and biodiversity, which in many cases results in the creation of wetland compensation sites through restoration or preservation. The NSW 1996 Wetlands Management Policy states that compensatory mitigation will be required in cases of wetland destruction that is in the public interest. In Canada, there is no co-ordinated federal policy regarding wetlands permitting and mitigation, and only four out of the 10 provinces have a mitigation policy that expresses a sequence of steps and compensation guidelines (Rubec and Hanson 2009). Spain has a biodiversity policy, also applying to wetlands, of "*no perdida neta*"[9], and draft NNL policies have also been developed in Hong Kong and Rwanda (R. Gardner, personal communication). China has no single statute that is directed solely at wetland protection, and while a system of reserves exists, some protective regulations have been enacted. However, there is no permit system and so nothing comparable to a formal mitigation procedure (Wen et al. 2005). The European Union Habitats Directive (EUHD), which emerged in the early 1990s, contains a kind of alternatives analysis in Sections 6(3) and 6(4), and has been applied (along with the Birds Directive) to regulate approximately 131 million acres of wetlands, more wetland area than remains in the continental US (Crooks et al. 2001).

However, the EUHD is characteristic of many global policy efforts that include wetland mitigation in that it is aimed primarily at *habitat*, not *wetlands* per se. To the extent that wetlands are valuable habitat, they may be protected (or mitigation required for impacts to them) by the EUHD and by laws such as those in Australia. However, the EUHD (much like Ramsar) requires the selection of candidate sites recognized for their significant habitat value rather than blanket application to a class of ecosystemic features. The specific federalist history of the United States, in which the federal government was given control over waters as a way of preventing inter-state trade wars, has created a policy environment where "waters" (a class to which many wetlands belong) are a legal object and a specific target of federal protection and mitigation in a way that they cannot easily be in the rest of the world.

Conclusion

Over the 35-year history of the Section 404 program, the USEPA and the ACOE have made great strides in developing and refining the program's mitigation requirements and associated policy. The federal resource agencies have supported dozens of small and large-scale evaluations of the third step in the sequence, *compensation*, to help them understand its strengths and weaknesses. Based on the lessons learned and recommendations from these studies, the agencies have generated comprehensive compensatory mitigation regulations[10] and over a half dozen national guidance documents[11] designed to elevate the success rate of compensation. A motivated focus on improving the effectiveness of avoidance and minimization could yield similar tangible results, but there has been almost no work carried out

---

[9] No net loss

[10] See Final Compensatory Mitigation Rule at: http://www.epa.gov/wetlandsmitigation/.

[11] See the National Compensatory Mitigation Guidance List: http://www.mitigationactionplan.gov/links.html.

on these subjects from either policy or ecological perspectives. This is an increasingly glaring omission. As Houck (1989, p. 838) reminds us, compensation "is a measure of last, not first, resort. Until this principle is actually implemented by permit review staffs, the concept of [compensation] will continue to wag the dog, pointing it away from those hard and necessary decisions that will avoid wetlands loss." To judge by the comments received on the 2008 compensation rule, it appears the public agrees with the NRC, which defined avoidance as "the first and most desirable of the sequencing steps in wetland mitigation" (National Research Council 2001, p. 299). We look forward to the important work that will aid in the effective implementation of all three steps in the mitigation process.

Acknowledgments The authors gratefully recognize the invaluable guidance and contributions of Bill Kruczynski of the United States Environmental Protection Agency, Region 4, Matt Schweisberg of the United States Environmental Protection Agency, Region 1, David Olson of Army Corps of Engineers, Regulatory Headquarters, former United States Environmental Protection Agency HQ Wetlands Division Directors Dave Davis and John Meaghar, and former United States Environmental Protection Agency HQ Wetlands Regulatory Branch Chief John Goodin. This research was conducted in part with assistance from a fellowship from the Oak Ridge Institute for Science and Education.

References

Brown JD (1989) Wetlands mitigation: US Fish and Wildlife Service policy and perspectives. In: Sharitz RR, Gibbons JW (eds) Freshwater wetlands and wildlife. United States Department of Energy, Office of Scientific and Technical Information, Oak Ridge, Tennessee, pp 861–868

Brumbaugh RW, Reppert R (1994) National wetland mitigation banking study: first phase report. Report IWR 94-WMB-4. United States Army Corps of Engineers, Institute for Water Resources, Alexandria, Virginia, p 80

Ciupek RB (1986) Protecting wetlands under Clean Water Act Section 404: EPA's conservative policy on mitigation. Natl Wetl Newsl 8:12–13

Conservation Foundation (1988) Protecting America's wetlands: an action agenda. The final report of the National Wetlands Policy Forum. The Conservation Foundation, Washington, p 69

Crooks S, Ledoux L, Fairbrass J (2001) No net loss the European Union way. Natl Wetl Newsl 23:1, 14–17

Downing D, Winer C, Wood L (2003) Navigating through Clean Water Act jurisdiction: a legal review. Wetlands 23:475–493

Environmental Law Institute (2007) Mitigation of impacts to fish and wildlife habitat: estimating costs and identifying opportunities. Environmental Law Institute, Washington, p 125

Frayer WE, Monahan TJ, Bowden DC, Graybill FA (1983) Status and trends of wetlands and deepwater habitats in the conterminous United States, 1950's to 1970's. United States Department of the Interior, Fish and Wildlife Service, Washington, p 31

Gardner RC (2000) Money for nothing? The rise of wetland fee mitigation. Va Environ Law J 19:1–56

Gardner RC (2003) Rehabilitating nature: a comparative review of legal mechanisms that encourage wetland restoration efforts. Cathol Univ Law Rev 52:573

Gosselink JG, Odum EP, Pope RM (1974) The value of the tidal marsh. LSU-SG-74-03. Center for Wetland Resources, Louisiana State University, Baton Rouge, Louisiana, p 30

Hall JR (1988) A perspective on influencing the Corps of Engineers. In: Kusler JA (ed) Mitigation of impacts and losses. Association of State Wetland Managers, Berne, pp 60–65

Houck OA (1989) Hard choices: the analysis of alternatives under section 404 of the Clean Water Act and similar environmental laws. Univ Colo Law Rev 60:773–840

Kantor RA, Charette DJ (1988) Origin, evolution, and results of New Jersey's wetlands mitigation policy. In: Kusler JA (ed) Mitigation of impacts and losses. Association of State Wetland Managers, Berne, pp 103–105

Kelly P (1989) Memorandum: permit elevation, plantation landing resort, Inc. Signed 21 April, 1989. United States Army Corps of Engineers, Washington, p 15. http://www.epa.gov/owow/wetlands/pdf/PlantationLandingRGL.pdf

Kruczynski WL (1990) Mitigation and the section 404 program: a perspective. In: Kusler JA, Kentula ME (eds) Wetlands creation and restoration: the status of the science. Island Press, Washington, pp 549–554

LaRoe ET (1978) Mitigation: a concept for wetland restoration. In: Montanari JH, Kusler JA (eds) Proceedings of the National Wetland Protection Symposium, Reston. United States Fish and Wildlife Service, United States Department of the Interior, Washington, pp 221–224, 6–8 June 1977

LaRoe ET (1986) Wetland habitat mitigation: an historical overview. Natl Wetl Newsl 8:8–10

Mack JJ, Micacchion M (2006) An ecological assessment of Ohio mitigation banks: vegetation, amphibians, hydrology and soils. Ohio Environmental Protection Agency, Technical Report WET/2006-1. Ohio Environmental Protection Agency, Division of Surface Water, Wetland Ecology Group, Columbus, Ohio, p 106

National Research Council (1995) Wetlands: characteristics and boundaries. National Academy Press, Washington, p 328

National Research Council (2001) Compensating for wetland losses under the Clean Water Act. National Academy Press, Washington, p 348

Race MS, Christie DR (1982) Coastal zone development: mitigation, marsh creation, and decision-making. Environ Manag 6:317–328

Race MS, Fonseca MS (1996) Fixing compensatory mitigation: what will it take? Ecol Appl 6:94–101

Reiss KC, Hernandez E, Brown MT (2007) An evaluation of the effectiveness of mitigation banking in Florida: ecological success and compliance with permit criteria. Florida Department of Environmental Protection, Tallahassee, Florida, p 146. http://www.dep.state.fl.us/water/wetlands/docs/mitigation/Final_Report.pdf

Robertson MM (2006) Emerging ecosystem service markets: trends in a decade of entrepreneurial wetland banking. Front Ecol Environ 6:297–302

Robertson MM (2007) Discovering price in all the wrong places: commodity definition and price under neoliberal environmental policy. Antipode 39:500–526

Rubec CDA, Hanson A (2009) Wetland mitigation and compensation: Canadian experience. Wetl Ecol Manag 17:3–14

Ruhl JB, Salzman J (2006) The effects of wetland mitigation banking on people. Natl Wetl Newsl 28:1, 9–14

Shabman LA, Scodari P, King DM (1994) National wetland mitigation banking study: expanding opportunities for successful mitigation: the private credit market alternative. Report IWR 94-WMB-3. United States Army Corps of Engineers, Institute for Water Resources, Alexandria, Virginia, p 76

Soileau DM (1984) Final report on the Tenneco Laterre corporation mitigation banking proposal, Terrebonne Parish, Louisiana. United States Fish and Wildlife Service, Division of Ecological Services, Lafayette, Louisiana, p 23

Spieles D (2005) Vegetation development in created, restored, and enhanced mitigation wetland banks of the United States. Wetlands 25:51–63

United States Army Corps of Engineers (1973) Permits for activities in navigable waters or ocean waters. Fed Regist 38:12217–12230

United States Army Corps of Engineers and United States Environmental Protection Agency (1990) Memorandum of agreement between the Department of the Army and the Environmental Protection Agency: the determination of mitigation under the Clean Water Act Section 404(b)(1) Guidelines. Signed February 6, 1990. Washington. http://www.wetlands.com/fed/moafe90.htm

United States Army Corps of Engineers and United States Environmental Protection Agency (2002) Guidance on compensatory mitigation projects under the Corps Regulatory Program pursuant to Section 404 of the Clean Water Act and Section 10 of the Rivers and Harbors Act of 1899. Regulatory Guidance Letter 02-2, Issued December 24, 2002. Washington. http://www.usace.army.mil/CECW/Documents/cecwo/reg/rgls/RGL2-02.pdf

United States Army Corps of Engineers and United States Environmental Protection Agency (2008) Compensatory mitigation for losses of aquatic resources; final rule. Fed Regist 73:19594–19705

United States Army Corps of Engineers, United States Environmental Protection Agency, United States Fish and Wildlife Service, National Marine Fisheries Service and Natural Resources Conservation Service (1995) Federal guidance for the establishment, use and operation of mitigation banks. Fed Regist 60:58605–58614

United States Environmental Protection Agency (1975) Part 230—navigable waters: discharge of dredged or fill material. Fed Regist 40:41291–41298

United States Environmental Protection Agency (1980) Guidelines for specification of disposal sites for dredged or fill material. Fed Regist 45:85336–85357

United States Environmental Protection Agency (1986) Final determination of the assistant administrator for external affairs concerning the Sweeden's swamp site in Attleboro, Massachusetts pursuant to section 404(c) of the Clean Water Act. Signed 13 May, 1986. p 59. http://www.epa.gov/owow/wetlands/regs/404c.html. Accessed 8 April 2007

United States Fish and Wildlife Service (1981) United States Fish and Wildlife Service mitigation policy; notice of final policy. Fed Regist 46:7644–7663

United States Fish and Wildlife Service (1983) Interim guidance on mitigation banking. Ecological services instructional memorandum no. 80. Unites States Fish and Wildlife Service, Washington

United States Government Printing Office (1989) Congressional record of the 101st Congress, vol 135, part 20 (11/8/89 to 11/15/89). Government Printing Office, Washington

Webb JW, Landin MC, Allen HH (1986) Approaches and techniques for wetlands development and restoration of dredged material disposal sites. In: Kusler JA, Quammen ML, Brooks RP (eds) Mitigation of impacts and losses. Association of State Wetland Managers, Berne, pp 132–134

Wen Y, Hou F, Hazenberg G (2005). Institution, legislation and policy analysis of China's wetland protection. For Stud China 7:55–60

Wilkinson J, Thompson J (2006) 2005 Status report on compensatory mitigation in the United States. Environmental Law Institute, Washington, p 110

Wood LD (2004) Don't be misled: CWA jurisdiction extends to all non-navigable tributaries of the traditional navigable waters and to their adjacent wetlands (a response to the Virginia Albrecht/Stephen Nicklesburg ELR article, to the fifth circuit's decision *in R. Needham*, and to the Supreme Court's dicta in *SWANCC*). Environ Law Rep 34:10187–10217

Yocom TG, Leidy RA, Morris CA (1989) Wetlands protection through impact avoidance: a discussion of the 404(b)(1) alternatives analysis. Wetlands 9:283–297

Chapter 11
The Ramsar Convention

Royal C. Gardner and Nick C. Davidson

Abstract This chapter examines the Ramsar Convention on Wetlands, the global intergovernmental treaty that promotes wetland conservation worldwide. When one is studying or seeking to protect a particular wetland, it is important to look beyond the wetland's delineated borders. As discussed in Chap. 1, the health of a wetland is influenced by its placement in the landscape, the ecosystem services it provides, and the activities that occur within its watershed (e.g., development, agriculture). Focusing solely on the wetland site may result in missing the bigger picture. Similarly, when studying wetland policies, it is instructive to look beyond domestic regimes (i.e., national and local laws and policies) and consider global wetland policies. Parties to the Ramsar Convention address this through three main pillars of implementation: the 'wise use' of all wetlands, the designation and management of Wetlands of International Importance (Ramsar Sites), and international cooperation on management of shared resources and sharing of knowledge and information.

Treaty Basics

A treaty is a written agreement between two or more countries that is governed by international law. Such an agreement might be formally called a treaty, convention, protocol, or other name, but the name is not critical. The important point is that the countries that are parties to the agreement intend to be bound by its provisions. A treaty between two countries is sometimes referred to as a bilateral agreement; a treaty with many parties is often called a multilateral agreement. The Ramsar Convention, with 160 parties as of December 2010, is a multilateral agreement. Because of its subject matter, the Ramsar Convention is also known as a *multilateral environmental agreement,* or MEA. Other examples of MEAs include the Kyoto Proto-

R. C. Gardner (✉)
Stetson University College of Law, 1401 61st Street South, Gulfport, FL 33707, USA
e-mail: gardner@law.stetson.edu

B. A. LePage (ed.), *Wetlands,*
DOI 10.1007/978-94-007-0551-7_11, © Springer Science+Business Media B.V. 2011

col, which deals with greenhouse gas emissions, the United Nations Convention on Biological Diversity (CBD), the Convention on International Trade in Endangered Species (CITES), and the Convention on Migratory Species (CMS).

The text of a treaty is developed through negotiations. The need and potential for an international governmental agreement on wetlands was recognized in the early 1960s by a number of non-governmental organizations (NGOs), governments, and wetland and waterbird ecologists, in the face of increasing concerns about losses of wetland habitat and associated declines in waterfowl populations. After over eight years of discussions, meetings, and negotiations, in 1971 governmental representatives from 18 countries (with observers from other countries, intergovernmental organizations (IGOs), and NGOs also present) met in the Iranian city of Ramsar, on the coast of the Caspian Sea, to seek to finalize the terms of the treaty. Thus, "Ramsar" is not an acronym; the Convention takes its name from the location in which it was negotiated just like the Geneva Convention or Kyoto Protocol. On February 2, 1971, the treaty (Convention on Wetlands of International Importance Especially as Waterfowl Habitat 1971) was concluded (i.e., the negotiators agreed on the final form of the text) and it was opened for signature. By its terms, the Ramsar Convention would enter into force (become effective) once seven countries agreed to become parties. The Ramsar Convention entered into force in December 1975. For a full account of the development and early history of the Convention, see Matthews (1993).

The Convention text establishes several procedural options for a country to become a Ramsar party. A country can join the treaty either by a notification of accession, signature of the Convention subject to ratification (followed by ratification), or signature without reservation of ratification. Each country becoming a Ramsar party can choose which procedure to follow, according to its own ratification procedures. In the United States, for example, the U.S. Senate must ratify a treaty, approving it by at least a two-thirds vote. The United States became a Ramsar party in 1987 during the Reagan administration after the U.S. Senate ratified the Ramsar Convention (Ramsar Convention Secretariat 2010). Whichever procedure is followed, the notification of joining the Convention must be sent to the United Nations Educational, Scientific and Cultural Organization, which is the Convention's legal depository organization, and must be accompanied by the designation of the country's first Wetland of International Importance (Ramsar site)—more information about which is below. The Convention then enters into force in that country four months after accession or ratification.

Duties Under the Ramsar Convention

The Ramsar Convention places upon its parties three primary obligations, which are sometimes referred to as the pillars of Ramsar: employing the "wise use" approach to wetlands; designating and conserving at least one site as a Wetland of International Importance; and international co-operation.

The Wise Use of Wetlands

Article 3 of the Ramsar Convention calls on Ramsar parties to "formulate and implement their planning so as to promote … as far as possible the wise use of wetlands in their territory." This requirement applies to all wetlands and water resources under a party's jurisdiction. The parties have defined "wise use" of wetlands to mean "the maintenance of their ecological character, achieved through the implementation of ecosystem approaches, within the context of sustainable development." The Ramsar Convention's concept of wise use is entirely consistent with the notion of sustainable development adopted at the Earth Summit in Rio de Janeiro two decades later. (Wise use in the Ramsar context should not be confused with the wise use movement, which is largely an anti-regulatory group, in the western United States.)

Ramsar parties fulfill their wise use obligations through various mechanisms at different levels of management (Ramsar Convention Secretariat 2007b; Resolution IX.1 2005). A party is expected to develop national wetland legislation, regulations, and/or policies. In the United States, examples include the federal regulatory program under the Clean Water Act (CWA) administered by the United States Army Corps of Engineers (ACOE) and the United States Environmental Protection Agency (USEPA). But it also includes non-regulatory approaches such as the Wetlands Reserve Program run by the Natural Resources Conservation Service (NRCS). Moreover, a wise use approach to wetland management entails knowledge about the ecosystem services provided by wetlands, data that can be provided through an inventory and monitoring program. Wetland research, training, education, and public participation are important components as well.

At the site level, the wise use provisions mean that wetlands should be managed through ecosystem-based approaches at the landscape/seascape scale, with often the watershed (catchment) forming the basis for implementation. Special attention should be paid to integrating wetland management with coastal zone and river basin plans. The parties should strive to maintain the ecological functions of individual sites, and the Ramsar community recognizes (as does this book) that effective wetland management requires an interdisciplinary effort involving biology, economics, law and policy, and the social sciences. The involvement of the local communities is also critical to the wise use of a site, whether the community uses the site itself or surrounding areas. Depending on the site, wise use can also involve restoration projects.

The List of Wetlands of International Importance

Article 2 of the Ramsar Convention states that each Ramsar party "shall designate suitable wetlands within its territory for inclusion in a List of Wetlands of International Importance." These Wetlands of International Importance are also known as Ramsar sites, for brevity. The Ramsar Convention provides some guidance on which wetlands qualify for this honor. Such sites "should be selected for the List on account of their international significance in terms of ecology, botany, zoology, limnology or hydrology." The Ramsar parties have established nine criteria for identifying a Ramsar site (Ramsar Convention Secretariat 2005). While a site need only meet one criterion to qualify, in practice Ramsar parties often designate sites that satisfy multiple criteria. The nine criteria are divided into two groups, one based on wetland types and the other related to different aspects of wetland biodiversity at ecological community and species levels (Fig. 11.1).

Group A. Sites containing representative, rare or unique wetland types

Criterion 1: A wetland should be considered internationally important if it contains a representative, rare, or unique example of a natural or near-natural wetland type found within the appropriate biogeographic region.

Group B. Sites of international importance for conserving biological diversity

Criteria based on species and ecological communities

Criterion 2: A wetland should be considered internationally important if it supports vulnerable, endangered, or critically endangered species or threatened ecological communities.

Criterion 3: A wetland should be considered internationally important if it supports populations of plant and/or animal species important for maintaining the biological diversity of a particular biogeographic region.

Criterion 4: A wetland should be considered internationally important if it supports plant and/or animal species at a critical stage in their life cycles, or provides refuge during adverse conditions.

Specific criteria based on waterbirds

Criterion 5: A wetland should be considered internationally important if it regularly supports 20,000 or more waterbirds.

Criterion 6: A wetland should be considered internationally important if it regularly supports 1% of the individuals in a population of one species or subspecies of waterbird.

Specific criteria based on fish

Criterion 7: A wetland should be considered internationally important if it supports a significant proportion of indigenous fish subspecies, species or families, life-history stages, species interactions and/or populations that are representative of wetland benefits and/or values and thereby contributes to global biological diversity.

Criterion 8: A wetland should be considered internationally important if it is an important source of food for fishes, spawning ground, nursery and/or migration path on which fish stocks, either within the wetland or elsewhere, depend.

Specific criteria based on other taxa

Criterion 9: A wetland should be considered internationally important if it regularly supports 1% of the individuals in a population of one species or subspecies of wetland-dependent non-avian animal species.

Fig. 11.1 Criteria for identifying and designating wetlands of international importance. (Ramsar Convention Secretariat 2005)

Each Ramsar party has its own internal process to determine which site or sites should be designated. For example, in the United States, the Director of the United States Fish and Wildlife Service (USFWS) makes the designation decision, and the process has a scientific and a political component (United States Fish and Wildlife Service 2008). A nominator must complete the Ramsar Information Sheet (RIS), which contains important data such as a proposed site's boundaries, its historic and current ecological condition, and threats to its ecological integrity. The nominator must also obtain the consent of the site's landowners and letters of support from the state's natural resource agency and a member of Congress. The nomination packet is then circulated within the government for comment. If the USFWS Director approves and designates the site, the RIS is then forwarded to the Ramsar Secretariat in Switzerland. In the United Kingdom, a proposed site must first be declared under domestic legislation (generally as a Site of Special Scientific Interest [SSSI]), and then those wetland sites that also qualify as internationally important are prepared for Ramsar site designation. In European Union countries the obligations under European law to declare Special Protection Areas (SPAs) for birds and Special Areas of Conservation (SAC) for habitats (the Natura 2000 network) are closely linked to wetland areas also being designated as Ramsar sites. Indeed, some of the criteria for identification of SPAs derive directly from the Ramsar waterbird criteria. In contrast, elsewhere and more often in the developing world, Ramsar designation can be the first formal recognition of the importance of a wetland.

Ramsar Secretariat staff review each RIS, and if it is complete and the information provided confirms that the site satisfies the criteria, the Secretariat then places the site on the List of Wetlands of International Importance. Figure 11.2 provides the locations of the 29 Ramsar sites designated by the United States. As of December 2010, the List contained 1,913 sites worldwide, covering nearly 187 million ha worldwide.

Once a site is listed, Article 3 of the Ramsar Convention imposes several duties upon the Ramsar party. Chief among them is the obligation to "promote the conservation" of the Ramsar site. Furthermore, a Ramsar party must notify the Ramsar Secretariat if the site's ecological character "has changed, is changing, or is likely to change as the result of technological developments, pollution, or other human interference." Note that the listing does not affect the territorial sovereignty of the site, nor does it place the site off-limits to human activity. Indeed, the preamble to the Ramsar Convention notes the interdependence of people and the environment, and emphasizes the ecosystem services that wetlands provide, calling them a "resource of great economic, cultural, scientific, and recreational value."

International Co-operation

Article 5 of the Ramsar Convention calls on the parties to cooperate regarding wetland matters. It requires that the parties consult with each other about treaty obligations, especially where they share wetlands, water systems, or migratory waterbird species. Article 5 also provides that the parties shall try to co-ordinate and support wetland conservation policies and regulations. This duty of international co-opera-

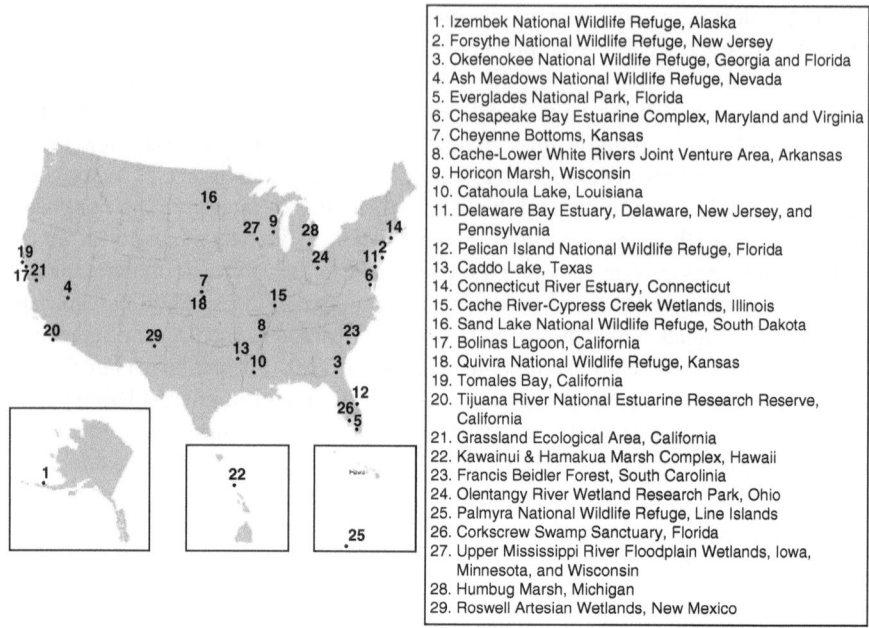

1. Izembek National Wildlife Refuge, Alaska
2. Forsythe National Wildlife Refuge, New Jersey
3. Okefenokee National Wildlife Refuge, Georgia and Florida
4. Ash Meadows National Wildlife Refuge, Nevada
5. Everglades National Park, Florida
6. Chesapeake Bay Estuarine Complex, Maryland and Virginia
7. Cheyenne Bottoms, Kansas
8. Cache-Lower White Rivers Joint Venture Area, Arkansas
9. Horicon Marsh, Wisconsin
10. Catahoula Lake, Louisiana
11. Delaware Bay Estuary, Delaware, New Jersey, and
    Pennsylvania
12. Pelican Island National Wildlife Refuge, Florida
13. Caddo Lake, Texas
14. Connecticut River Estuary, Connecticut
15. Cache River-Cypress Creek Wetlands, Illinois
16. Sand Lake National Wildlife Refuge, South Dakota
17. Bolinas Lagoon, California
18. Quivira National Wildlife Refuge, Kansas
19. Tomales Bay, California
20. Tijuana River National Estuarine Research Reserve,
    California
21. Grassland Ecological Area, California
22. Kawainui & Hamakua Marsh Complex, Hawaii
23. Francis Beidler Forest, South Carolinia
24. Olentangy River Wetland Research Park, Ohio
25. Palmyra National Wildlife Refuge, Line Islands
26. Corkscrew Swamp Sanctuary, Florida
27. Upper Mississippi River Floodplain Wetlands, Iowa,
    Minnesota, and Wisconsin
28. Humbug Marsh, Michigan
29. Roswell Artesian Wetlands, New Mexico

Fig. 11.2 United States Ramsar sites. http://www.ramsarcommittee.us/Ramsar_Sites_Brochure.pdf

tion can take many forms, including: participating in Ramsar conferences, meetings and regional initiatives; working to protect transboundary sites and species conservation; working together to maintain the ecological character of all Ramsar sites; and providing financial and technical assistance to other parties.

Parties are expected to attend and participate in the triennial Conference of the Contracting Parties, which is the Convention's policy and decision-making body. This gathering offers an opportunity for the parties to share information and discuss how the Convention is being implemented, as well as approving budgets and adopting decisions on future implementation actions and priorities (details about this conference are provided below).

Parties may also cooperate through regional wetland initiatives. An example is the Mediterranean Wetlands Initiative (known as MedWet), which co-ordinates activities in the Mediterranean Basin with the goal of halting and reversing wetland loss and degradation (The Mediterranean Initiative 2009). Twenty-five parties, as well as the Palestine Authority, UN agencies, and NGOs, are involved in this effort. Other more recent examples of regional initiatives recognized within the framework of the Convention include the Ramsar Regional Center for Training and Research on Wetlands in the Western Hemisphere (CREHO) in Panama, the West African Coastal Zone Wetlands Network, the Ramsar Regional Training Center for Training and Research on Wetlands in Western and Central Asia located in Iran, and the Regional Strategy for the Conservation and Wise Use of High Andean Wetlands.

Parties also satisfy the duty of international co-operation when they work to-gether to conserve transboundary sites. A number of states have designated their Ramsar sites as transboundary sites, which means that "an ecologically coherent wetland extends across national borders and the Ramsar authorities on both or all sides of the border have formally agreed to collaborate in its management" (Ramsar Convention Secretariat 2006). For example, the Floodplains of the Morava-Dyje-Danube Confluence are a trilateral site consisting of designations by Austria, the Czech Republic, and the Slovak Republic.

Article 5 specifically mentions that parties are to co-ordinate approaches to conserve wetland flora and fauna. There are many such efforts with respect to the conservation of migratory waterbirds through flyway programs, such as the North American Waterfowl Management Plan (NAWMP) implemented by Can-ada, Mexico, and the United States. NAWMP seeks to increase shared waterfowl populations through habitat improvements. Since its inception in 1986, $4.5 billion has been spent to restore, enhance, and/or protect 15.7 million acres of waterfowl habitat (United States Fish and Wildlife Service 2009). On other flyways there are a range of statutory and non-statutory migratory waterbird initiatives, such as the African-Eurasian Migratory Waterbird Agreement (AEWA), the East Asian-Australasian Flyway Initiative, and the Western Hemisphere Shorebird Reserve Network (WHSRN).

Provisions regarding financial and technical assistance from developed countries have become a common feature in multilateral environmental agreements. Although the Ramsar Convention does not expressly require developed countries to provide wetland-related assistance to developing countries, many Ramsar parties do so, and these assistance programs are another facet of international co-operation. For ex-ample, the Small Grants Fund for Wetland Conservation and Wise Use, which is administered by the Ramsar Secretariat, operates through voluntary contributions from more than a dozen developed country Ramsar parties. From 1991 through 2008 the Small Grants Fund has financed 277 projects in 108 countries (Ramsar Convention Secretariat 2009). Individual parties also work with the Ramsar Secre-tariat to promote wetland conservation on a regional basis, as is the case with the United States-funded Wetlands for the Future, which provides assistance in the Ca-ribbean and Latin America, and the Swiss Grant for Africa, which finances Ramsar-related initiatives in Africa.

Ramsar Players and Process

As with any treaty regime, there are a host of different entities that work together to promote the Ramsar Convention's objectives (Fig. 11.3). The primary decision-and policy-making body is the *Conference of the Contracting Parties,* known as the COP, which is made up of government representatives from the parties, as well as observers from non-party states, IGOs, NGOs, and others. The COP meets every three years to assess how the Convention is being implemented and to consider res-

Fig. 11.3 The relationships between various Ramsar entities at national and global scales. (Ramsar Secretariat 2007a). *STRP:* Scientific & Technical Review Panel, *CEPA:* Communications, Education, Participation and Awareness, *NFP:* National Focal Point

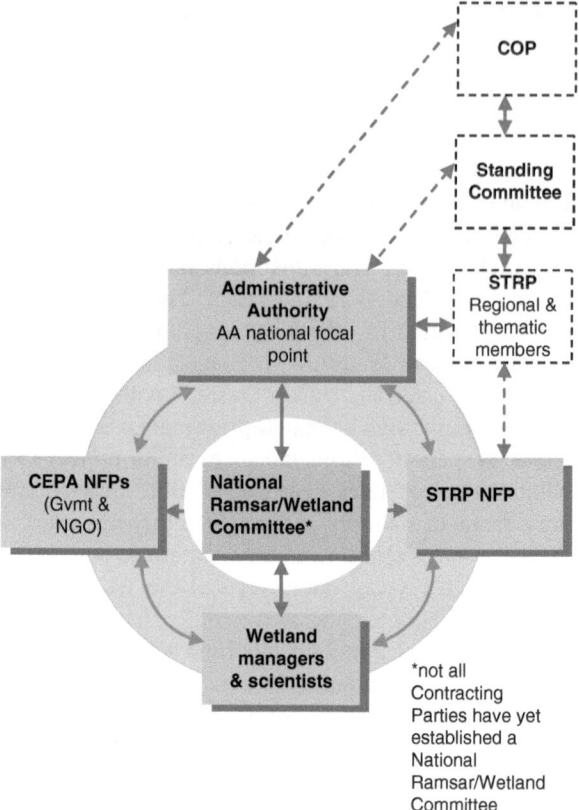

olutions and recommendations related to wetland conservation. It is a long-upheld Ramsar tradition that COP decisions are made by consensus.

The *Standing Committee,* which was established by a COP resolution, serves as the COP's inter-sessional governance body. The COP elects the members of the Standing Committee on a regional basis, with the number of representative member Parties for each Ramsar region being determined by the number of parties in that region. Ordinarily, the Standing Committee meets once a year, usually at the Secretariat Headquarters in Gland, Switzerland.

The *Ramsar Secretariat* is responsible for coordinating the day-to-day activities of the Ramsar Convention. The Secretariat is led by the Secretary General, who is appointed by the Standing Committee and reports to it. The staff includes a Deputy Secretary General (the Convention's senior scientific, technical, and policy advisor) and senior regional advisors (for Africa, the Americas, Asia-Oceania, and Europe). The Secretariat has formal responsibilities for many of the activities discussed in this chapter, such as maintaining the List of Wetlands of International Importance, organizing and reporting to Ramsar COPs and other meetings, providing support to Ramsar parties, and helping administer wetland assistance programs. The Secretariat also hosts a comprehensive website that provides a wealth of wetland-related information and guidance (www.ramsar.org). Of particular note is the Ramsar Toolkit, a series of seventeen handbooks that provide a thematically organized range of guidance on the wise use and conservation of wetlands adopted by the various COPs (Ramsar Convention Secretariat 2007b).

The *Scientific and Technical Review Panel* (STRP), established by COP5 Resolution 5.5 in 1993 (Resolution 5.5 1993), is a subsidiary Convention body that provides scientific and technical advice to the COP, Standing Committee and Secretariat, and reports to the Standing Committee. The STRP's latest *modus operandi* was adopted in COP10 Resolution X.9 (Resolution X.9 2008). Individual STRP thematic and regional members are appointed as wetland experts in their own right by the STRP Oversight Committee, following nominations from the parties, STRP national focal points, STRP members, and observer organizations. The primary outputs of the STRP are draft technical resolutions that go to the COP for review and adoption, technical implementation guidelines for the parties that are typically annexed to resolutions considered at the COP, and more detailed technical reviews and reports that are published by the Secretariat as Ramsar Technical Reports (which do not go to the COP for approval and are available on the Ramsar website).

Each Ramsar party must designate an *Administrative Authority,* which may be a government ministry, department, or other national agency that is responsible for implementing Ramsar domestically. The Administrative Authority is the national focal point for Ramsar and, as such, is the method by which the Secretariat communicates with a party. Ramsar parties are also expected to designate several experts to serve as liaison with wetland managers, scientists, and communications experts. There are two national focal points for *Communication, Education, Participation, and Awareness* (*CEPA*) activities (one governmental, the other non-governmental), as well as a national focal point for STRP matters.

A COP recommendation and the Ramsar strategic plan encourage parties to establish *National Ramsar Committees* (Recommendation 5.7 1993; Resolution X.1 2008). These committees may help develop national wetland policies, offer guidance on the management of Ramsar sites, provide input about adding new sites to the Ramsar List, and review and advise on the implementation of the resolutions and recommendations adopted by the COP. There is no prescribed form for a National Ramsar Committee. Some consist of only government representatives, while others, as is the case in the United States, may be more NGO-based (United States National Ramsar Committee 2009). Regardless of the structure, the point of the committee is to involve at a national level the relevant stakeholders from government, scientific and technical institutions, local communities, NGOs, and the private sector so as to enhance the national implementation of the Ramsar Convention.

The Legal Effect of the Ramsar Convention

A fundamental principle of international law is *pacta sunt servada*—promises or agreements must be kept—and this principle applies to Ramsar parties and the Ramsar Convention. But, like most MEAs, many of the Convention's duties are written in a manner that affords the parties flexibility and discretion in how to implement them. Accordingly, the Ramsar Convention is often considered "soft law," espousing aspirational goals, but not dictating binding legal obligations.

The view that the Ramsar Convention is soft law does not mean, however, that it is without effect on a party's domestic laws and internal decision-making processes. In some countries, such as Australia, the Ramsar Convention and the designation of sites have been used to augment the national government's authority over environmental issues (Finlay-Jones 1997). In other countries, Ramsar guidance has prompted the development of comprehensive wetland policies and strategic plans (Bowman 2002; Gardner 2003). A case out of the Netherlands Antilles illustrates that Ramsar resolutions can even affect site-level permitting decisions (Verschuuren 2008).

In 2006, the government of Bonaire, the local authority of the Dutch territory of Netherlands Antilles, authorized a resort to be built adjacent to Het Lac, a Ramsar site. The Governor-General annulled the granting of the permit because the environmental impact assessment failed to satisfy the commitments in Resolution VIII.9 adopted by the parties in 2002 (Resolution VIII.9 2002). Bonaire appealed to the Dutch Crown, contending that Ramsar resolutions are not legally binding. In November 2007 Queen Beatrix issued a royal decree upholding the annulment reasoning that "resolutions, decisions and guidelines accepted unanimously by the Conference of Parties to the Convention, of which the Netherlands is a signatory, must be considered part of the national obligations under the Convention" (Newton 2007).

Not all parties accept the legal theory that consensus-approved resolutions at conferences of the parties create binding domestic obligations. There is case law to the contrary in the United States in the context of the Montreal Protocol on Substances that Deplete the Ozone Layer, for example (National Resource Defense Council (NRDC) v. EPA 2006). Regardless of whether the resolutions are considered binding law or an expression of goals, they influence the actions of the parties and other Ramsar stakeholders.

The Benefits of the Ramsar Designation

If the Ramsar Convention is not regulatory in nature, then what benefits does it offer beyond serving as a forum to exchange views and information? Recently, surveys of Ramsar site managers were conducted in the United States, Canada, and Africa (Gardner and Connolly 2007; Lynch-Stewart and Associates 2008; Gardner et al. 2009). Each survey yielded similar findings: the designation of a site as a Wetland of International Importance was more than a mere honor; the status offered tangible benefits. The type and level of benefits varied from site to site.

Ramsar designation was found to contribute to the protection and management of the sites. In Canada, two-thirds of site managers reported that Ramsar designation helped maintain the ecological character of the site by communicating the site's value to the public, influencing landuse and development activities, and focusing attention on long-term stewardship. In the United States, Ramsar designation was instrumental in expanding the buffers or boundaries of some protected areas (e.g.,

Edwin B. Forsythe National Wildlife Refuge in New Jersey), as well as encouraging partnerships in watershed conservation efforts such as the Cache River Wetlands Joint Venture Partnership in Illinois. The international designation has also assisted some sites with restoration work (e.g., Blackwater National Wildlife Refuge in Maryland, part of the Chesapeake Bay complex). Moreover, even though the Ramsar Convention is viewed as a non-regulatory entity in the United States, it is invoked in regulatory settings. For example, at zoning hearings near Tomales Bay, California, people reminded decision-makers that their landuse actions may have an effect on an internationally significant site. The designation of a site can play an educational role about wetlands and increase a local community's support of the site.

Ramsar designation can also help increase financial support for a site. Over 80% of the U.S. respondents stated that Ramsar designation assisted with grant applications and other funding requests. Approximately 65% of the African sites surveyed also noted increased funding opportunities as a benefit, through the Ramsar Small Grants Fund and other mechanisms. Local groups that support Ramsar sites have also found that the international status aids with grant applications.

Because Ramsar sites must meet specified ecological criteria, it is not surprising that the confirmation of a site's significance can result in increased ecotourism. In Africa, ecotourism can be a component of poverty alleviation. Furthermore, many site managers report an increased scientific interest post-designation. The data resulting from these studies "should likely reinforce the other benefits offered by Ramsar designation. Increased knowledge should help educate the public and decision makers about the importance of a site, thus leading to increased support for its protection" (Gardner and Connolly 2007).

Conclusion: The Evolution and Future of the Ramsar Convention

As one of the earliest MEAs, the Ramsar Convention has been increasingly recognized as providing a supportive and flexible mechanism for addressing the many challenges faced by governments worldwide in achieving wetland conservation and wise use.

The text of the Convention was inspirational for its time in the way it strongly recognized the interdependence of people and wetlands, and the critical roles that wetlands play in the hydrological cycle and sustainable water management. Although environmental language usage has changed since that of the 1970s, the Convention has kept pace with such changes, by updating (at COP9 in 2005) its definitions of the key terms such as wetland "wise use" and "ecological character" to reflect modern concepts and terms such as sustainable use, ecosystem approaches and ecosystem services (Resolution IX.1 Annex A).

As Matthews (1993) describes, an initial focus of Convention implementation was the designation of Ramsar sites, with a first priority for designating areas of international importance for migratory waterbirds. This proved to be an effective means of 'getting Convention implementation going' and indeed the ever-growing network of Ramsar sites continues to be a flagship mechanism for the Convention, and is now by far the largest network of protected areas (in the broad sense of the term) in the world. Nevertheless, there remain significant gaps in the Ramsar site network, including areas for waterbirds. Many recognizably internationally important wetlands are not yet formally designated, and a number of these are under threat, continue to deteriorate or are permitted by governments and others to be destroyed through conversion to other landuses. Beyond designation, a major continuing challenge is to sustainably manage Ramsar sites so as to maintain their ecological character in the face of many on- and off-site pressures.

However, the Convention's scope has always been much broader than just Ramsar site designation. In particular, the wise use provisions are the core tenet of Convention implementation. Delivering the wise use of all wetlands has proved even more challenging for countries to achieve than designating further Ramsar sites, since it requires co-ordinated cross-sectoral ecosystem-based action to achieve, and through this achievement to maintain the huge value to people of the ecosystem services that wetlands can, and still do, provide (e.g., Costanza et al. 1997). Underlying this challenge is the need for much better data and information for decision-makers on the values of multiple wetland services. In turn, decision-makers must better understand and recognize these values in landuse planning, since conversion of naturally functioning wetlands to other, often single-sector uses has been shown to often reduce their overall value (e.g., Balmford et al. 2002).

In its early days, much Convention implementation focused on conserving just the wetlands themselves, rather than cross-sectorally delivering wise use in its landscape context. This attitude has progressively shifted to a much wider recognition that wetland conservation in most parts of the world cannot be delivered unless the economic and ecological value of wetlands to other sectors (including people from local communities to the private sector) is recognized, and all such sectors are engaged in maintaining wetlands and their services. Key progress was not taken until Ramsar COP6 in 1996, 25 years after the signing of the Convention, when parties adopted Resolution VI.23 on "Ramsar and water" (Resolution VI.23 1996). This resolution stressed, as was written into the original Convention text itself, that not only do wetlands depend for their survival on the presence of water, but also that wetlands play a pivotal role in managing and processing water for the many uses that people need it for, such as agricultural irrigation, energy generation, drinking, and sanitation. Since then the Convention has developed and adopted a major suite of wetlands and water-related guidance to help parties in working between those responsible for Ramsar wetland implementation and those managing water and its uses at catchment (basin) scales.

Yet despite this, and other COP resolutions since 1996 addressing the linkage between wetlands, water and other sectors such as fisheries, agriculture, and human health, decision-making in the context of current imperatives for economic growth are leading to continuing deterioration and destruction of coastal and inland wetlands. Indeed, the Millennium Ecosystem Assessment (MEA) reported that inland and coastal wetlands were being lost at a faster rate than any other ecosystem (Finlayson et al. 2005), and recent, largely species-based, assessments from the 2010 Biodiversity Indicators Partnership (Butchart et al. 2010) indicate that rates of loss in the 1990s/2000s have accelerated rather than reduced.

For the immediate future, the key priority must be to bring to the attention of all parts of government and civil society the full importance of wetlands for our future survival, particularly in relation to adapting and responding to our rapidly changing climate. To that end, importantly Ramsar COP10 in 2008 adopted the "Changwon Declaration on wetlands and human well-being" (Resolution X.3 2008), designed to speak not to the converted (wetland conservation practitioners), but to the less-converted in these other sectors whose future business success depends just as much on maintaining wetlands and their services.

So after almost 40 years, how effective has the Convention been in wetland conservation and wise use? On the face of it, with losses of wetlands and wetland-dependent species continuing largely unabated, it could be seen as not having been effective. But it is better to ask if there is evidence as to how much worse the state of wetlands would have been without the many and inspirational efforts by people and organizations worldwide to implement the Convention. Current work by the Convention's STRP on "ecological outcome-oriented indicators of the effectiveness of the Convention," requested by parties at COP8, is finding evidence that the national status of wetlands is affected positively if the Convention is being substantively implemented. This is particularly the case if a party has a National Wetland Policy or an equivalent national enabling framework in place and is also undertaking a range of other national implementation activities.

Yet many countries lack implementation capacity and/or the cross-sectoral political will to ensure such landscape-scale collaborative implementation. Enhancing the capacity and understanding of the value of wetlands to all sectors of society must be the key priority for the future work under the Convention   and to ensure that those responsible as Ramsar Administrative Authorities have the necessary expertise, capacity, and national mandates to take this message to those others in national governance who need it for their own sustainable futures, through maintaining and restoring the world's wetlands for people and wildlife.

References

Balmford A, Bruner A, Cooper P, Costanza R, Farber S, Green RE, Jenkins M, Jerreriss P, Jessamy V, Madden J, Munro K, Myers N, Naeem S, Paavola J, Rayment M, Rosendo S, Roughgarden J, Trumper K, Turner RK (2002) Economic reasons for conserving wild nature. Science 297:950–953

Bowman M (2002) The Ramsar convention on wetlands: has it made a difference? In: Stokke OS, Thommessen ØB (eds) Yearbook of international co-operation on environment and development. Earthscan, London, pp 61–68

Butchart SHM, Walpole M, Collen B, van Strien A, Scharlemann JPW, Almond REA, Baillie JEM, Bomhard B, Brown C, Bruno J, Carpenter KE, Carr GM, Chanson J, Chenery AM, Csirke J, Davidson NC, Dentener F, Foster M, Galli A, Galloway JN, Genovesi P, Gregory RD, Hockings M, Kapos V, Lamarque J-F, Leverington F, Loh J, McGeoch MA, McRae L, Minasyan A, Morcillo MH, Oldfield TEE, Pauly D, Quader S, Revenga C, Sauer JR, Skolnik B, Spear D, Stanwell-Smith D, Stuart SN, Symes A, Tierney M, Tyrrell TD, Vié J-C, Watson R (2010) Global biodiversity: indicators of recent declines. Science. doi:10.1126/science.1187512 (Published online 29 Apr 2010)

Convention on Wetlands of International Importance Especially as Waterfowl Habitat. Ramsar (Iran) (1971) United Nations Treaty series No. 14583. As amended by the Paris Protocol, Dec 3, 1982, and Regina Amendments. http://www.ramsar.org/cda/ramsar/display/main/main.jsp?z n=ramsar&cp=1-31-38^20671_4000_0__. Accessed 28 May 1987 (2 Feb)

Costanza R, d'Arge R, de Groot R, Farberk S, Grasso M, Hannon B, Limburg K, Naeem S, O'Neill RV, Paruelo J, Raskin RG, Sutton P, van den Belt M (1997) The value of the world's ecosystem services and natural capital. Nature 397:253–260

Finlay-Jones J (1997) Aspects of wetland law and policy in Australia. Wetl Ecol Manag 5:37–54

Finlayson CM, d'Cruz R, Davidson NC (2005) Ecosystems and human well-being: wetlands and water. Synthesis. Millennium Ecosystem Assessment. World Resources Institute, Washington, p 68

Gardner RC (2003) Rehabilitating nature: a comparative review of legal mechanisms that encourage wetland restoration efforts. Cathol Univ Law Rev 52:573–620

Gardner RC, Connolly KD (2007) The Ramsar Convention on wetlands: assessment of international designations within the United States. Environ Law Rep 37:10089–10113

Gardner RC, Connolly KD, Bamba A (2009) African wetlands of international importance: assessment of benefits associated with designations under the Ramsar Convention. Georget Int Environ Law Rev 21:257–294

Lynch-Stewart and Associates (2008) Wetlands of international importance (Ramsar sites) in Canada: survey of Ramsar site managers 2007, final report. Ottawa, Ontario, Canada

Matthews GVT (1993) The Ramsar Convention on wetlands: its history and development. Ramsar Convention Bureau, Gland, Switzerland, p 120

National Resource Defense Council (2006) National Resource Defense Council v. EPA, 464 F.3d 1 (D.C. Cir. 2006)

Newton EC (2007) Annulment of decisions for building near Ramsar site on Bonaire was justified. http://www.ramsar.org/cda/ramsar/display/main/main.jsp?zn=ramsar&cp=1-26-76^18583_ 4000_0__. Accessed 11 Dec 2007

Ramsar Convention Secretariat (2005) The criteria for identifying wetlands of international importance. http://www.ramsar.org/cda/ramsar/display/main/main.jsp?zn=ramsar&cp=1-36-55%5E20740_4000_0__

Ramsar Convention Secretariat (2006) The Ramsar Convention manual: a guide to the convention on wetlands (Ramsar, Iran, 1971), 4th edn. Ramsar Convention Secretariat, Gland, Switzerland, p 114. http://www.ramsar.org/cda/en/ramsar-pubs-manual-ramsar-convention-21302/ main/ramsar/1-30-35^21302_4000_0__

Ramsar Convention Secretariat (2007a) Delivering the Ramsar Convention in your country, national focal points and their roles. http://www.ramsar.org/pdf/about/about_nfp_2007_e1.pdf

Ramsar Convention Secretariat (2007b) Ramsar handbooks for the wise use of wetlands, 3rd edn. Ramsar Convention Secretariat, Gland, Switzerland, p 26. http://www.ramsar.org/pdf/lib/lib_ handbooks2006_e01.pdf

Ramsar Convention Secretariat (2009) Working for wetlands: the small grants fund for wetland conservation and wise use—SFG, project portfolio 2009. http://www.ramsar.org/pdf/sgf/sgf_ portfolio_2009.pdf. Accessed 4 Nov 2009

Ramsar Convention Secretariat (2010) Contracting parties in order of their accession. http:// www.ramsar.org/cda/en/ramsar-about-parties-contracting-parties-in-20715/main/ramsar/1-36-123%5E20715_4000_0__. Accessed 31 Dec 2010

Recommendation 5.7, National Committees (1993) Recommendations of the 5th meeting of the conference of the contracting parties. Kushiro, Japan, 9–16 June 1993

Resolution 5.5, Establishment of a scientific and technical review panel (1993) Resolutions of the 5th meeting of the conference of the contracting parties. Kushiro, Japan, 9–16 June 1993

Resolution VI.23, Ramsar and water (1996) Resolutions of the 6th meeting of the conference of the contracting parties. Brisbane, Australia, 19–27 March 1996

Resolution VIII.9 (2002) Guidelines for incorporating biodiversity-related issues into environmental impact assessment legislation and/or processes and in strategic environmental assessment adopted by the Convention on Biological Diversity (CBD), and their relevance to the Ramsar Convention. 2002. Resolutions of the 8th meeting of the conference of the contracting parties. Valencia, Spain, 18–26 Nov 2002

Resolution IX.1 (2005) Additional scientific and technical guidance for implementing the Ramsar wise use concept. Resolutions of the 9th meeting of the conference of the contracting parties. Kampala, Uganda, 8–15 Nov 2005

Resolution X.1 (2008) The Ramsar strategic plan 2009–2015. Resolutions of the 10th meeting of the conference of the contracting parties. Changwon, Republic of Korea, 28 Oct to 4 Nov 2008

Resolution X.3 (2008) The Changwon declaration on human well-being and wetlands. Resolutions of the 10th meeting of the conference of the contracting parties. Changwon, Republic of Korea, 28 Oct to 4 Nov 2008

Resolution X.9 (2008) Refinements to the modus operandi of the Scientific & Technical Review Panel (STRP). Resolutions of the 10th meeting of the conference of the contracting parties. Changwon, Republic of Korea, 28 Oct to 4 Nov 2008

The Mediterranean Initiative of the Ramsar Convention on Wetlands (2009) About MedWet. http://www.medwet.org/medwetnew/en/02.ABOUT/02.0.about_medwet.html. Accessed 4 Nov 2009

United States Fish and Wildlife Service (2008) U.S. Ramsar site nomination and designation factsheet. Arlington, Virginia. http://www.ramsarcommittee.us/documents/Designationfactsheet200800509.pdf

United States Fish and Wildlife Service, Division of Bird Habitat Conservation (2009) North American waterfowl management plan. Accomplishments. http://www.fws.gov/birdhabitat/nawmp/index.shtm. Accessed 12 Jan 2009

United States National Ramsar Committee (2009) Home page. http://www.ramsarcommittee.us/index.asp (4 Nov)

Verschuuren J (2008) Ramsar soft law is not soft at all: discussion of the 2007 decision by the Netherlands Crown on the Lac Ramsar site on the island of Bonaire. Milieu en Recht 35:28–34. http://www.ramsar.org/pdf/wurc/wurc_verschuuren_bonaire.pdf

Chapter 12
Managing Wetlands for Multifunctional Benefits

Robert J. McInnes

Abstract Humans have sought to manage and exploit wetlands from the beginning of recorded history. From the early modifications of the marshlands of Mesopotamia to the draining of the English Fens, the management of wetlands has always reflected the major societal and economic drivers of the times. Early efforts at conservation management focussed primarily on waterbirds, but over the latter part of the twentieth Century there has been a shift of emphasis towards towards 'wise use' and an understanding of how ecosystem processes and functions deliver important benefits to human society. The recognition that the natural environment provides societal benefits is crucial to the future conservation of wetlands. Building on the Ecosystem Approach, the adoption of a multidisciplinary approach to the management of wetlands, which extends beyond biological and hydrological regimes into social and economic science, is considered essential if improved decision-making is to be achieved in order to deliver multifuncitonal benefits from wetland management.

Paleoecological, archaeological, and historical data demonstrate that humans have sought to exploit and manage wetland environments, and the important services that they provided since the dawn of civilization (Denham et al. 2003). While the magnitude of change imposed on the environment in pre-history pales in comparison with events over the last century, there is evidence that the implications of even the earliest attempts at natural resource management have imprinted a legacy on wetlands that is manifest today. For instance, humans have been disturbing coastal wetlands since they first learned how to fish. Data drawn from the paleoecological and archaeological records have demonstrated that early changes to coastal marine ecosystems, such as oyster harvesting or alterations in sedimentation rates due to evolving land management practices, increased the sensitivity of near-shore

R. J. McInnes (✉)
Bioscan (UK) Ltd, The Old Parlour, Little Baldon Farm, Little Baldon,
Oxford, OX44 9PU, UK
e-mail: robmcinnes@bioscanuk.com

B. A. LePage (ed.), *Wetlands,*
DOI 10.1007/978-94-007-0551-7_12, © Springer Science+Business Media B.V. 2011

wetlands to subsequent disturbance and thus pre-conditioned the collapse in global fisheries that we are witnessing today (Jackson et al. 2001).

Often historical changes in wetland management are less insidious and, superficially, more dramatic. One of the earliest documented accounts of wetland management developed around 8,000 years Before Present (BP) when Neolithic farmers began to migrate into the floodplain forming a 'fertile crescent' bordering the Tigris and Euphrates Rivers, known in ancient days as Mesopotamia (derived from the Greek meaning literally 'between the rivers') (Algaze 2008). This plain was dissected by river channels with bountiful stocks of fish and replenished by alluvial silt laid down routinely by seasonal flooding. In the words of V. Gordon Childe, one of the great archaeological synthesizers of the twentieth Century, the area held considerable potential for emerging agricultural practices where 'arable land had literally to be created out of a chaos of swamps and sand banks by a "separation" of land from water; the swamps ... drained; the floods controlled; and life-giving waters led to the rainless desert by artificial canals' (McInnes 2008). The local population overcame a range of environmental challenges by cooperative social effort. Between 5,500 and 5,100 years BP, the foundations were laid in this region for a type of social order and economy that is now considered the cradle of Western Civilization (Thesiger 1964).

The Mesopotamian marshes between the Tigris and Euphrates Rivers were once the largest wetlands in southwest Asia and covered more than 15,000 km², an area nearly twice the size of the original extent of the Florida Everglades. For generations these wetlands were managed in a sustainable way and were home to between 300,000 and 500,000 indigenous Marsh Arabs (Young 1977; Coast 2002). However, as a result of a systematic plan by Saddam Hussein's administration to ditch, dike, and drain the marshes of southern Iraq, less than 10% of the area remained as functioning marshland by the year 2000 (Richardson and Hussain 2006). Despite recent efforts to reverse this loss, in the space of a generation, traditional sustainable wetland management has been replaced by a destructive and potentially irreversible management regime (Richardson et al. 2005).

On a smaller scale to the Mesopotamian Marshes, the Fens of eastern England once covered an area of approximately 5,000 km². During the seventeenth Century the Earl of Bedford employed the pre-eminent Dutch drainage engineer Cornelius Vermuyden to undertake what was probably the most challenging scheme of his career and for which he was eventually knighted. His task was to drain the 'Great Fen' in Cambridgeshire. This operation was financed by the Earl of Bedford, and the 'Great Fen' became known thereafter as Bedford Level (Harris 1953). The principal engineering achievements resulted in the construction of two major channels, the Old Bedford River and the Forty Foot Drain, and a drainage channel known as the New Bedford River.

Vermuyden's intensive management of the wetlands of the Fens reduced, but did not eradicate, the potential of the region to flood, despite the initial success of the reclamation. However, the drying of the land caused the carbon-rich peat soil to mineralize and shrink rapidly due to oxidation, lowering the land below the level of the drainage channels and rivers. By the end of the seventeenth Century, much of the land reclaimed by Vermuyden was again waterlogged and would remain so until the arrival of steam-powered pumps in the nineteenth Century. The drainage and intensification of agriculture across this area of former wetland has continued into the twenty-first Century. However, there are now attempts to reconcile the competing demands on the Fens as a simultaneous provider of food, rural livelihoods, and environmental goods and services through a more sustainable management regime and the restoration of extensive wetland areas (Morris et al. 2000; Acreman et al. 2007; Bowley 2007).

The management of wetlands will reflect the major societal and economic drivers of the times. The ambitions of the Earl of Bedford to create a productive agricultural landscape are reflected elsewhere such as in the United States. At the time of European settlement in the early 1600s, the area that was to become the conterminous United States had approximately 894,355 km^2 of wetlands (Dahl and Alford 1996). Wetland drainage began with the permanent settlement of Colonial America in the 1600s and has continued into the twenty-first Century. By the 1960s, most political, financial, and institutional incentives to drain or destroy wetlands were in place. The federal government encouraged land drainage and wetland destruction through a variety of legislative and policy instruments. For example, the Watershed Protection and Flood Prevention Act (United States Congress Public Law 1954) directly and indirectly required an increase in the drainage of wetlands near flood-control projects (Erickson 1979). The federal government directly subsidized or facilitated wetland losses through its many public-works projects, technical practices, and cost-shared drainage programs administered by the United States Department of Agriculture (USDA) (Erickson 1979). Perversely, the drainage of wetlands through tile and open-ditch systems was considered a conservation practice under the Agriculture Conservation Program, whose policies caused wetland losses averaging 2,226 km^2 each year from the mid-1950s to the mid-1970s (Office of Technology Assessment 1984). It has been estimated that the intensification of agriculture was responsible for more than 80% of these wetland losses (Frayer et al. 1983).

The management of wetlands down through the ages appropriately illustrates the contemporaneous needs and values of society. The establishment of the roots of Western Civilization in the Mesopotamian Marshes demanded the involvement of a multiplicity of disciplines including proto-engineers, agronomists, geologists, and administrators. This multidisciplinary approach to the management of wetlands has been replicated across the world as humans have sought to manage wet areas for their collective benefit.

The Origins of Wetland Conservation Management

During the first decades of the twentieth Century wetlands began to be assigned a degree of protection from increasing threats imposed by pollution, drainage, and modification. The National Association of Audubon Societies, named after the American ornithologist John James Audubon (1780–1851), was formed in 1902 and directed considerable effort to protect the great (*Ardea alba*) and snowy egrets (*Egretta thula*) in Florida. Similar endeavours in the Netherlands saw the Vereniging tot Behoud van Natuurmonumenten (Society for the Preservation of Nature Reserves) purchase the Naardermeer Bird Reserve in 1906 to protect it from becoming a waste dump for the city of Amsterdam.

Progressively over the first quarter of the twentieth Century international nature protection moved away from 'usefulness for man' and began to develop management prescriptions that represented wider perspectives. The International Committee for Bird Protection (ICBP) was formed in 1922 (changing its name to the International Council for Bird Protection in 1958) with the expressed conviction 'that wild bird-life is of great importance in the world in helping to preserve the balance between species, which nature is constantly seeking to adjust, that birds have a great importance to science, exercise a great aesthetic influence on all right-minded people, and are of great value to mankind as food, as destroyers of rodents and injurious insects, and as incentives for reasonable field-sports' (Bezemer 1974).

In the United States, attention frequently focused on ensuring the protection and appropriate management of wetlands for wetland birds. The National Wildlife Refuge system was formed in 1942, and almost all initial refuge acquisitions were wetland systems. The acquisition of many of these areas was facilitated through the migratory bird stamp funds generated from waterfowl hunters (Salyer and Gillet 1964).

In 1947, the ICBP founded the International Wildfowl Research Institute, renamed the International Wildfowl Research Bureau (IWRB) in 1954. In 1962, the IWRB and the ICBP, together with the International Union for Conservation (IUCN), initiated a conference at which it was agreed to compile a list of wetlands of international significance and, on the basis of that list, to arrive at a convention for the protection and management of those areas. Following on from the 1962 conference, a series of technical meetings and negotiations culminated in the 'Convention on Wetlands of International Importance Especially as Waterfowl Habitat' (better known as the Ramsar Convention, see Gardner and Davidson 2011).

The Ramsar Convention imposes obligations upon its parties that should result in the better management of wetland ecosystems: either through the embracing the principles of 'wise use'; by designating at least one internationally important site; and through international cooperation. In December 2010 there were 160 signatories to the Ramsar Convention with 1,913 sites designated covering an area of nearly 1.87 million km^2. Wetlands remain the only group of ecosystems to have their own international convention.

Before, and in parallel with the development and subsequent implementation of the Ramsar Convention, concerns over declines in fish and wildfowl populations stimulated the wider conservation movement (Carp 1980). The early management efforts to conserve fish and wildlife were often aimed at simply protecting conservation lands from development (Kusler 1983). Much of the focus of wetland management, and restoration in particular, continued to be on abating and reversing the loss and decline of wildlife populations and particularly waterbirds (Davidson and Evans 1987; Benstead 2000).

A Shift in Emphasis

Despite the concept of 'wise use', wetland functions and ecosystem processes were not generally or widely understood or considered important until relatively recently. In the United States, until the later part of the last century ecosystem processes influencing wetlands were not generally included in formal curricula or postgraduate training for most wetland managers (Euliss et al. 2008). The failure to recognise linkages between wetlands and wider spatial relationships beyond habitats for fauna, such as with landscape aesthetics (Minich 2011) or global carbon budgets (Middleton 2011), has contributed to continued wetland loss (Maltby et al. 1996; Crooks et al. 2001). However, this single focus approach on protecting a preferential wetland resource without considering the broader functioning of the ecosystem has lead to conflicts and potentially negative impacts on wetlands (Euliss et al. 2008).

In some quarters this approach continues today. The rarity and decline of the Eurasian bittern (*Botaurus stellaris*) in the United Kingdom has prompted large-scale wetland restoration and more recently, wetland creation projects (Gilbert et al. 2005). While the conservation of a single species is an admirable and important goal, in recent times there has been a growing dichotomy between the preservation of a single-species and broader, ecosystem-based approaches to conservation (Sergio et al. 2003). Often considerable capital resource is invested in the acquisition, design, and restoration of wetlands, with a focus on a single or charismatic species, as is the case with the Eurasian bittern in the United Kingdom. While serendipitous benefits may accrue (McInnes 2007) the invisibility of, or worse still, the turning of a blind eye to, broader spatial relationships can perpetuate wetland loss and degradation (Turner et al. 2000).

Since the 1970s there has been a burgeoning recognition that wetlands provide many important services to human society, but are at the same time they remain ecologically sensitive systems (Nyman 2011; Maltby 1986; Millennium Ecosystem Assessment 2005). However, it has been argued that despite the broader understanding that wetlands are part of our natural capital and wealth creation potential (Costanza et al. 1997) continued wetland loss has been caused by: (1) the public nature of many wetland services or products; (2) user externalities (a cost or benefit not transmitted through prices) imposed on other stakeholders; and (3) policy intervention failures that are due to a lack of consistency among policies enacted across different sectors (Turner et al. 2003). There is strength in the argument that good stewardship and sustainable management of wetland resources requires a paradigm shift (Maltby 2009a) and the need for multidisciplinary approaches (LePage 2011) if we are to overcome these barriers.

The Ecosystem Approach and Wetland Management

Such a new paradigm should embrace the ecosystem approach as a strategy for the integrated management of land, water, and living resources that promotes conservation and sustainable use in an equitable way (Convention on Biological Diversity

2010). The term 'Ecosystem Approach' was first applied in a policy context at the Earth Summit in Rio de Janeiro in 1992, where it was adopted as an underpinning concept of the Convention on Biological Diversity. It is now an integral component of environmental policy. For example it has been endorsed by the World Summit on Sustainable development in Johannesburg in 2002, and is implicit in the European Water Framework Directive (Commission of the European Communities 2000), the approach implemented to halt the loss of biodiversity by 2010 as agreed in Gothenburg by the European Union Heads of Government and the Ramsar Convention (Beaumont et al. 2007).

Such a holistic, interdisciplinary perspective should place wetlands centrally in the implementation of the ecosystem approach (Fig. 12.1). Linkages between society and the natural environment on the one hand, and environmental management on the other, should be made explicit in order to achieve sustainable use and management of wetlands that is valuable for, and valued by, society as a whole (Hahn

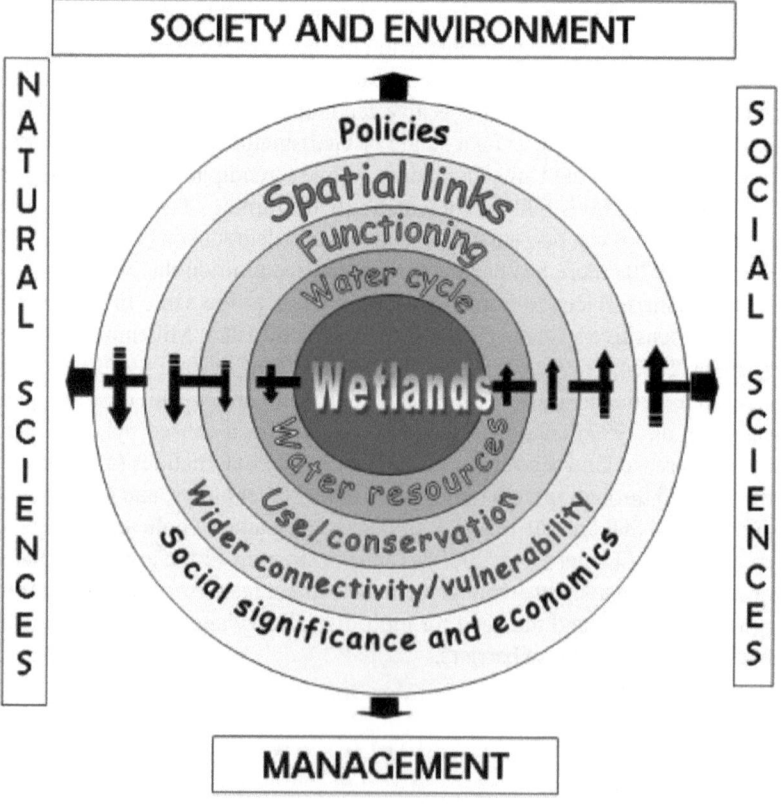

Fig. 12.1 A new paradigm with wetlands at the centre of a new multidisciplinary perspective of the priority concerns of society, linking natural and social sciences. (Reproduced from Maltby 2009a)

et al. 2006). These linkages should be realized by way of the roles wetlands have in the water cycle, ecosystem functioning, spatial relationships, and government policies, and feed into management of water resources, use and conservation of wetland resources, connectivity and vulnerability in the landscape, social significance, and the economic values of wetlands in providing ecosystem services (Maltby 2009a).

Advocates of the ecosystem approach argue that because knowledge about the complexity and inter-connectedness of ecosystems is imperfect, management needs to be adaptive and include a means of learning about ecosystem dynamics through the implementation of policy experiments (Dale et al. 2000). It is further argued that the evolving engagement with a diversity of stakeholders often forms part of the learning process of ecosystem management (Mitlin and Thompson 1995; Walker et al. 2002). The societal response to the acidification of the Lake Racken area in western Sweden illustrates how an environmental crisis can act as a catalyst for adaptive ecosystem management involving collective action and social learning.

The threat of acidification of Swedish freshwater ecosystems has been well documented since the late 1960s (Dickson 1978; Korsman 1999). During the 1970s in the Lake Racken area there was a growing concern that acidification was compromising water quality and degrading fisheries. The originator of this concern was a local resident who was also a technician at the municipal water works at Lake Racken with the appropriate skills and access to equipment to catalogue changes in water quality. Observations of decreasing pH levels had begun in 1971 in the tributaries and small lakes of the Lake Racken catchment. Taking advantage of Swedish government funds that were available to formal liming groups, the technician mobilized other residents and formed a liming group that commenced liming of the lake. This local initiative preceded the national monitoring program by several years. The response of the local community reversed the acidification trend in the area and started a knowledge accumulating process with the objective of maintaining Lake Racken's desirable ecosystem state of a clear water lake with fish and associated ecosystem services (Olsson et al. 2004).

Comprising local landowners, the liming group self-organized into the Lake Racken Fishing Association in 1986. Since 1986, the Association has implemented measures to enhance brown trout (*Salmo trutta*) and noble crayfish (*Astacus astacus*) populations by restoring habitats and reducing threats within the Lake Racken catchment. In 1994, there was an important change in Swedish national legislation that devolved decision-making to local fishing associations regarding the use and management of fish and crayfish in inland freshwater bodies, with the exception of the three largest freshwater lakes in Sweden. Prior to this change, there was a requirement for local fishing associations to consult with the county administration board about any change in management practices. This legislative change generated increased flexibility and space for further self-organization, such as testing different management practices and rules at the local level, thereby tightening environmental feedback loops (Olsson et al. 2004).

The intelligence underpinning the ecosystem management implemented by members of the fishing association is a combination of scientific knowledge and local observation. The repository for the intelligence is within the social memory of the group. The management strategy has imparted knowledge within the group and this approach continues to draw on external sources for information in an enduring learning-by-doing process. For example, members of the association are aware of local temporal and spatial variations of acidification and how such variations are linked to water quality and organisms. Furthermore, they understand how short duration increases in acidity resulting from snowmelt affect crayfish physiology and reproduction and consequently implement targeted measures aimed at reducing the risk of such sudden changes in pH. The association also utilizes internal monitoring for fish and crayfish management. Members routinely monitor the pH, alkalinity, calcium levels, metal concentrations, and several indicator species including invertebrates and fish. Over time the fishing association has developed linkages with a social network of associations as well as strong links at municipal and county level. In the 1990s the fishing association was invited to participate in a joint policy initiative between Norway and Sweden with the objective of securing and enhancing the noble crayfish populations. The diversity of practices that have evolved as a part of the self-organizing process towards catchment-scale fisheries management is described in further detail by Olsson and Folke (2001). However, what is clear from this example is that the emergence of the adaptive co-management system from the liming group, to the fishing association, to the social network of associations, to linkages with municipal and county levels, and to the joint pan-national policy initiative has increased the likelihood of building social-ecological resilience in the area.

Application of this holistic approach can be made only by interdisciplinary collaboration and communication between and within the natural and social sciences and all participants (see Kaplan 2011). Wetland management strategies need to reflect this approach. However, a common misconception is that the ecosystem approach is an *ecosystems approach*. Examination of the literature clearly reveals that the ecosystem approach is not a set of guidelines for the management of various ecosystems, but a framework for thinking ecologically that results in actions that are based on holistic decision-making (Shepherd 2004; Laffoley et al. 2004). This framework for action links the multiple disciplines of *inter alia* biology, social sciences, and economics and aims to achieve a socially acceptable balance between nature conservation priorities, resource use, and the sharing of benefits. In Europe, for instance, there has been a movement away from addressing problems in isolation on land, in freshwaters, and in estuaries or the coastal zone to integrating these different geographical areas through an evolution from sectoral approaches to a holistic ecosystem approach to managing coastal and marine areas (Apitz et al. 2006).

The ecosystem approach was developed to bring clarity to the concept of sustainability, and it has been adopted as the primary framework for action under the Convention on Biological Diversity. The Convention has adopted twelve guiding principles and five points of operational guidance (Tables 12.1 and 12.2). Many of these principles are illustrated by the example of Lake Racken.

Table 12.1 The twelve principles of the ecosystem approach

1. The objectives of management of land, water and living resources are a matter of societal choice
2. Management should be decentralized to the lowest appropriate level
3. Ecosystem managers should consider the effects (actual or potential) of their activities on adjacent and other ecosystems
4. Need to understand and manage the ecosystem in an economic context
5. Conservation of ecosystem structure and function to provide ecosystem services should be a priority
6. Ecosystem must be managed within the limits of their functioning
7. The approach should be taken at the appropriate spatial and temporal scales
8. Process and objectives for ecosystem management should be set for the long term
9. Management must recognise that change is inevitable
10. Seek the appropriate balance between integration, conservation, and use of biodiversity
11. Decision-making should consider all forms of relevant information (scientific, indigenous, and local)
12. Involve all relevant sectors of society and scientific disciplines

Table 12.2 The five points of operational guidance for the ecosystem approach

1. Focus on the relationship and processes within the ecosystem
2. Enhance benefit sharing
3. Use adaptive management practices
4. Carry out management actions at the scale appropriate to the issue, with decentralization to the lowest level appropriate
5. Ensure intersectoral co-operation

Ecosystem Services and Wetland Management

While the ecosystem approach advocates that the objectives of the management of land, water, and living resources are a matter of societal choice, any decision needs to be based on a thorough understanding of the environmental limits that constrain ecological processes across a variety of scales. Despite the aspiration that the conservation of ecosystem structure and function to provide ecosystem services should be a priority, if a detailed understanding of which services an ecosystem can provide is absent, society is likely to make an ill-informed choice. Similarly, the sustainable use of ecosystem services is unlikely without improved understanding of the capacity of ecosystems to provide these services (Gunderson and Holling 2002).

The utilization of an approach around ecosystem services, as expressed in the fifth principle (Table 12.1), has the capacity to play a fundamental role in the multidisciplinary management of wetlands. While this is not an application of the ecosystem approach *per se,* the recognition that the natural environment provides benefits to human society is crucial to conserving wetlands. Since the publication of the Millennium Ecosystem Assessment (MEA) in 2005, there has been a growing emphasis

Table 12.3 Categories of ecosystem service and examples of related services. (Based on Millennium Ecosystem Assessment 2005)

Type of service	Service
Provisioning services	Food
	Fibre and fuel
	Genetic resources
	Bio-chemicals, natural medicines, etc.
	Fresh water
Regulating services	Climate regulation
	Water regulation
	Erosion regulation
	Natural hazard regulation
	Water purification and waste treatment
	Pollination
Cultural services	Spiritual and inspirational
	Recreation and ecotourism
	Aesthetic values
	Educational values
Supporting services	Soil formation
	Nutrient cycling

on identifying and valuing wetland ecosystem services (de Groot et al. 2006, Maltby 2009b). The MA developed a nomenclature that recognized four broad types of service, namely: *provisioning* services; *regulating* services; *cultural* services; and the *supporting* services that underpin the other three types. Each service possesses sub-categories. For instance *regulating services* include: climate regulation; water regulation; water purification and waste treatment; erosion regulation; and natural hazard regulation (Table 12.3).

Ecosystems services result from the interactions and processes within the ecosystem (de Groot et al. 2006). Wetlands are composed of a number of *biophysical structures* such as soil, water, and plant and animal species (Nyman 2011). The interactions among and within the biophysical structures that are present within a wetland ultimately result in ecosystem *processes,* such as denitrification, decomposition, or primary production. The interactions among and within these different components allow the wetland to perform certain ecological *functions* (de Groot 1992; Maltby et al. 1996; McInnes et al. 1998). The degree to which a wetland delivers ecosystem *services* depends on its functional properties (*e.g.,* biotic and abiotic components) and the relationship between and among ecological components and processes (Fig. 12.2). The use of functional wetland classifications can facilitate the understanding of these inter-relationships (Brinson 2011).

However, it is important to understand the societal and spatial context within which ecosystem services are operating or recognized. Figure 12.2 indicates that a wetland may have the capacity to provide one or several ecological functions (i.e., it possesses the appropriate biophysical structure) and supports the necessary processes that combine to produce these functions. However, it does not always follow

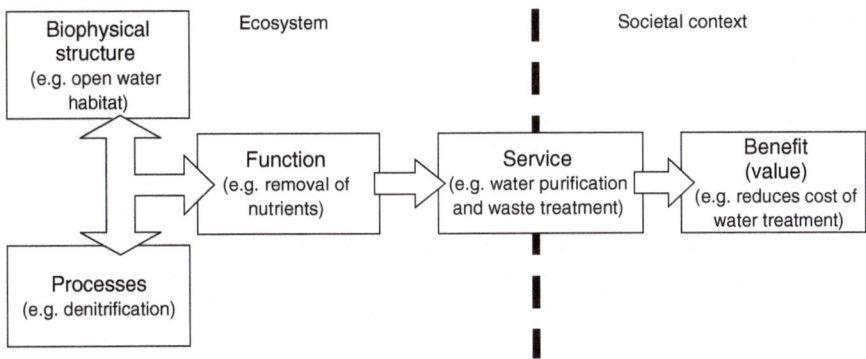

Fig. 12.2 The logic underlying the context of ecosystem services. (Adapted from Maltby et al. 1996; de Groot et al. 2006)

that society perceives a benefit (value) for the delivery of a particular function or service such as water purification or waste treatment. People and society value ecological functions differently, in different places, at different times (Haines-Young and Potschin 2007). Therefore, when defining what constitutes an ecosystem service, it is important to understand the broader multidisciplinary context that extends to include societal, spatial, and temporal issues.

In many quarters interest in the management of wetlands to provide ecosystem services has arisen. However, some concerns have been identified with such an approach. In the United States, commercial wetland mitigation banking has been developed collaboratively since the early 1990s as a way to offset wetland impacts. A key element in wetland mitigation banking is the aspiration to develop a market in privately-owned wetland ecosystem services, such as the regulation of flooding or the improvement of water quality, as a way of achieving the goals of the United States Clean Water Act of 1977. However, while this has proved attractive to practitioners and regulators, major obstacles have been identified including the problem of relying on ecological science to define the unit of trade (the ecosystem service) as well as the problem of aligning the somewhat independent relations of law, politics, markets, and ecosystems across an array of spatial scales (Robertson 2004). Similarly, work in Sweden on the use of wetlands to reduce the impact of agricultural activities on freshwater systems has shown that trying to maximize biodiversity and nutrient removal may not generate one-dimensional outcomes (Hansson et al. 2005). Results from a study of 32 constructed wetlands in southern Sweden show that a combination of the wetland features, namely shallow depth, large surface area, and high shoreline complexity are likely to provide a high biodiversity of birds, benthic invertebrates, and macrophytes and to have high nitrogen retention capabilities. Whereas to achieve more efficient phosphorus retention a small, deep wetland is likely to be required, but this will be less valuable in terms of biodiversity (Hansson et al. 2005).

An additional note of caution is required to ensure that sustainable wetland management is achieved. By emphasizing the economic value of ecosystem services it shifts the focus from fair, integrated, and well-informed management to understanding and quantifying (usually in economic terms) the services the natural environment provides for us, and then managing the environment so that the provision of these services is sustained for us over the long term (Pound 2009). In a study of the Willamette Basin in Oregon, a spatially explicit modelling tool was used to predict changes in ecosystem services, biodiversity conservation, and commodity production levels. The study estimated the social value of carbon sequestration (all sequestration, not just the portion of sequestration that would be eligible for trading in a carbon offset market), but concluded that whether the social value of carbon will decrease, increase, or remain constant in the future is uncertain (Nelson et al. 2009). The commoditisation of carbon, and especially the role that wetlands can play in sequestering carbon, is an extension of the historic conversion of nature into tradable commodities. Even in the emerging, embryonic carbon market, as with most markets, the price of carbon is mediated by supply and demand. If market demand is high, then the value of carbon sequestration as an ecosystem service may distort management strategies in preference of a high economic value services at the expense of other services that are not necessarily as profitable as carbon sequestration. Consequently without careful management decision-making there remains a risk that the silver bullet of ecosystem services may backfire, with high value wetland ecosystem services being promoted at the expense of conservation priorities (Vira and Adams 2009).

Taking Multidisciplinary Wetland Management Forward

When implemented with full regard to the underlying principles, the Convention on Biological Diversity ecosystem approach embraces the extensive elements of a multidisciplinary approach that extends beyond the biological or hydrological regimes into social and economic science. Pound (2009) has recommended that to take such an approach to wetland management forward, it will mean that we need to begin the process of trying to get to grips with the following:

- Undertaking and using 'systems thinking' approaches and techniques that start by considering the wetland system as a whole, rather than looking at bits of the system and trying to sum the parts. This includes identifying processes, linkages, and feedback mechanisms between the different parts of the system.
- Developing an increased understanding of the wetland system including: how to define wetland ecosystem(s); how the system functions; understanding ecosystem resilience; spatial and temporal scales; relationships with adjacent or linked ecosystems; and natural change.
- Seeing humans as intrinsic parts of the wetland system as we use, influence, and alter wetlands like other species and processes, albeit to a much greater extent. This includes: a better understanding of the human systems used in managing wetlands; a better understanding of the harvesting and use of the resources; and

how we effect and are affected by the ecosystems that are a part of our physical and social fabric.

• Improving decision-making processes. Essential to this issue will be environmental organizations learning to relinquish control and making decisions with other stakeholder rather than for them. This will require integrated, participatory, and effective stakeholder processes, which bring people with different types of knowledge and know-how together to better develop understanding, undertake principled negotiation, and make well-informed decisions.

Good examples exist where a truly integrated and multidisciplinary approach to wetland management has been adopted. The Integrated Wetland Assessment Toolkit developed by the IUCN (Springate-Baginski et al. 2009) seeks to set out a process for integrated wetland assessment to deliver sustainable wetland management. The toolkit provides methods that can be used to investigate the links between biodiversity, economics, and livelihoods in wetlands, and to identify and address potential conflicts of interest between conservation and development objectives. The integrated approach presented in the toolkit also enables practitioners to assess a wetland in terms of its combined biodiversity, economic, and livelihood values.

Similarly, the Millennium Ecosystem Assessment identified that the increase in the demand for agricultural provisioning services would intensify pressure on wetlands as a result of a burgeoning global population (Millennium Ecosystem Assessment 2005). The need to better understand the relationships between the sustainability of agricultural production and the management of other ecosystem services lead to the development of Guidelines on Agriculture and Wetlands Interactions (GAWI). The guidelines were developed as a co-operative venture between the Ramsar Convention, FAO, Wageningen University and Research Centre, International Water Management Institute, and Wetlands International and they have produced a supporting framework for the sustainable management of different types of wetland-agriculture interactions that have at their heart the need for multidisciplinary and participatory approaches to wetland management (Wood and van Halsema 2008).

In an attempt to seek social, economic, and environmental coherence among the issues associated with diffuse and point source pollution from agricultural wastewater, the Irish government has developed the 'Integrated Constructive Wetland' (ICW) initiative (Harrington and Ryder 2002). The ICW concept combines various approaches to water, land, and living resource management by explicitly integrating three objectives: water quality and water-quantity management, including flood hazard regulation; landscape-fit towards improving site aesthetic values; and enhancing biodiversity (Scholz et al. 2007). Based on the 're-animation' and creation of shallow, emergent-vegetated wetlands to intercept and treat water polluted from agricultural practices, the explicit integration of these cross discipline objectives is made with the expectation of achieving positive synergies that might not otherwise be achieved through traditional land or wetland management strategies.

While initially developed for the Annestown stream catchment (approximately 25 km^2) in south county Waterford, Ireland, the ICW concept has spread across the Republic of Ireland and is recognized as possessing the potential for universal application (Carty et al. 2008). Within the European Union, the appropriate application of its principles has relevance in contributing towards Member States achieving good chemical and ecological status for inland and coastal waters by 2015 as required under the Water Framework Directive (Commission of the European Communities 2000). In addition to the sustainable management of livestock wastewater research has demonstrated that ICW systems can also deliver a multiplicity of benefits without compromising the initial design objectives (Harrington and McInnes 2009). The ICW concept continues to be developed by applying the principles of adaptive management (Harrington et al. 2005) and has established an empathy and appreciation for wetlands within the local community that is widely recognized as being critical to secure sustainable use and protection of wetlands (Boyer and Polasky 2004).

Such initiatives demonstrate that the aspiration to apply multidisciplinary approaches to wetland management, which balance diverse interests in a sustainable and equitable manner, is feasible and achievable. The key challenge to wetland managers is to ensure that this approach continues to be embraced in order to protect and enhance wetlands.

References

Acreman MC, Fisher J, Stratford CJ, Mould DJ, Mountford JO (2007) Hydrological science and wetland restoration: some case studies from Europe. Hydrol Earth Syst Sci 11:158–169

Algaze G (2008) Ancient Mesopotamia at the dawn of civilization: the evolution of an urban landscape. Chicago University Press, Chicago, p 246

Apitz SE, Elliot M, Fountain M, Galloway TS (2006) European environmental management: moving to an ecosystem approach. Integr Environ Assess Manag 2:80–85

Beaumont NJ, Austen MC, Atkins JP, Burdon D, Degraer S, Dentinho TP, Derous S, Holm P, Horton T, van Ierland E (2007) Identification, definition and quantification of goods and services provided by marine biodiversity: implications for the ecosystem approach. Mar Pollut Bull 54:253–265

Benstead P (2000) Practical wetland restoration: some recent experiences of the Royal Society for the protection of birds. Landscape Res 25:394–398

Bezemer KWL (1974) The international council for bird preservation. Het Vogeljaar 22:611

Bowley A (2007) The great fen—a waterland for the future. Br Wildl 18:415–423

Boyer T, Polasky S (2004) Valuing urban wetlands: a review of non-market valuation studies. Wetlands 24:744–755

Brinson MM (2011) Classification of wetlands. In: LePage BA (ed) Wetlands—integrating multidisciplinary concepts. Springer, Dordrecht

Carp E (1980) A directory of wetlands of international importance in the western Palearctic wetlands. International Union for Conservation of Nature and Natural Resources (IUCN) and United Nations Environment Programme (UNEP), Gland, Switzerland, p 506

Carty A, Scholz M, Heal K, Gouriveau F, Mustafa A (2008) The universal design, operation and maintenance guidelines for Farm Constructed Wetlands (FCW) in temperate climates. Bioresour Technol 99:6780–6792

Coast E (2002) Demography of the Marsh Arabs. In: Nicholson E, Clark P (eds) The Iraqi marshlands: a human and environmental study. Politico's, London, pp 19–35

Commission of the European Communities (2000) Directive 2000/60/EC of the European Parliament and of the council establishing a framework for the community action in the field of water policy. http://www.tematea.org/?q=node/1312&PHPSESSID=d324807c78b71c5ec5c62 99c28f56d0a

Convention on Biological Diversity (2010) Operational guidance for application of the ecosystem approach. http://www.cbd.int/ecosystem/operational.shtml. Accessed 18 May 2010

Costanza R, d'Arge R, de Groot R, Farber S, Grasso M, Hannon B, Limburg K, Naeem S, O'Neill RV, Paruelo J, Raskin RG, Sutton P, Van den Belt M (1997) The value of the world's ecosystem services and natural capital. Nature 387:252–259

Crooks S, Turner RK, Pethig JS, Parry ML (2001) Managing catchment-coastal floodplains: the need for a UK water and wetlands policy. Centre for Social and Economic Research on the Global Environment (CSERGE) Working Paper PA 01-01, University of East Anglia, Norwich, p 21

Dahl TE, Alford GJ (1996) History of wetlands in the conterminous United States. United States Geological Survey, Water-Supply Paper 2425:19–26

Dale VH, Brown S, Haeuber RA, Hobbs NT, Huntly N, Naiman RJ, Riebsame WE, Turner MG, Valone TJ (2000) Ecological principles and guidelines for managing the use of land. Ecol Appl 10:639–670

Davidson NC, Evans PR (1987) Habitat restoration and creation: its role and potential in the conservation of waders. Wader Study Group Bulletin 49, Supplement/IWRB Special Publication 7:139–145

Denham TP, Haberle SG, Lentfer C, Fullagar R, Field J, Therin M, Porch N, Winsborough B (2003) Origins of agriculture at Kuk Swamp in the highlands of New Guinea. Science 301:189–193

Dickson W (1978) Some effects of the acidification of Swedish lakes. Verhandlungen Internationalen Vereinigung Theoretische Angewandte Limnologie 20:851–856

Erickson RE (1979) Federal programs influencing wetlands. Seventh Annual Michigan Landuse Policy Conference, Michigan State University, East Lansing, Michigan, p 246

Euliss NH Jr, Smith LM, Wilcox DA, Browne AB (2008) Linking ecosystem processes with wetland management goals: charting a course for a sustainable future. Wetlands 28:553–562

Frayer WE, Monahan TJ, Bowden DC, Graybill FA (1983) Status and trends of wetlands and deepwater habitats in the conterminous United States, 1950s to 1970s. Colorado State University, Fort Collins, p 31

Gardner RC, Davidson NC (2011) The Ramsar convention. In: LePage BA (ed) Wetlands integrating multidisciplinary concepts. Springer, Dordrecht

Gilbert G, Tyler GA, Dunn CJ, Smith KJ (2005) Nesting habitat selection by bitterns Botaurus stellaris in Britain and the implications for wetland management. Biol Conserv 124:547–553

de Groot RS (1992) Functions of nature: evaluation of nature in environmental planning, management and decision-making. Wolters Noordhoff BV, Groningen, p 345

de Groot RS, Stuip MAM, Finlayson CM, Davidson N (2006) Valuing wetlands: guidance for valuing the benefits derived from wetland ecosystem services. Ramsar Technical Report No. 3/ CBD Technical Series No. 27. Ramsar Convention Secretariat, Gland, p 46

Gunderson LH, Holling CS (2002) Panarchy: understanding transformations in human and natural systems. Island Press, Washington, p 508

Hahn T, Olsson P, Folke C, Johansson K (2006) Trust-building, knowledge generation and organizational innovations: the role of a bridging organization for adaptive co-management of a wetland landscape around Kristianstad, Sweden. Hum Ecol 34:573–592

Haines-Young R, Potschin M (2007) The ecosystem concept and identification of ecosystem goods and services in english policy context. Review Paper to Defra, London, UK. Project code NR0107, p 21

Hansson L-A, Broenmark C, Nilsson PA, Aabjoernsson K (2005) Conflicting demands on wetland ecosystem services: nutrient retention, biodiversity or both? Freshw Biol 50:705–714

Harrington R, McInnes RJ (2009) Integrated Constructed Wetlands (ICW) for livestock wastewater management. Bioresour Technol 100:5498–5505

Harrington R, Ryder C (2002) The use of integrated constructed wetlands in the management of farmyard runoff and waste water. In: Proceeding of the National Hydrology Seminar on Water Resources Management Sustainable Supply and Demand. The Irish National Committees of the IHP (International Hydrological Programme) and ICID (International Commission on Irrigation and Drainage), Tullamore, pp 55–63

Harrington R, Dunne EJ, Carroll P, Keohane J, Ryder C (2005) The concept, design and performance of integrated constructed wetlands for the treatment of farmyard dirty water. In: Dunne EJ, Reddy KR, Carton OT (eds) Nutrient management in agricultural watersheds: a wetlands solution. Wageningen Academic Publishers, Wageningen, pp 179–188

Harris LE (1953) Vermuyden and the fens. Cleaver Hume Press, London, p 168

Jackson JBC, Kirby MX, Berger WH, Bjorndal KA, Botsford LW, Bourque BJ, Bradbury RH, Cooke R, Erlandson J, Estes JA, Hughes TP, Kidwell S, Lange CB, Lenihan HS, Pandolfi JM, Peterson CH, Steneck RS, Tegner MJ, Warner RR (2001) Historical overfishing and the recent collapse of coastal ecosystems. Science 293:629–638

Kaplan R (2011) Acknowledging and benefitting from multiple perspectives. In: LePage BA (ed) Wetlands—integrating multidisciplinary concepts. Springer, Dordrecht

Korsman T (1999) Temporal and spatial trends of lake acidity in northern Sweden. J Paleolimnol 22:1–15

Kusler JA (1983) Our national wetland heritage—a protection guidebook. Environmental Law Institute, Washington, p 4

Laffoley D d'A, Maltby E, Vincent MA, Mee L, Dunn E, Gilliland P, Hamer JP, Mortimer D, Pound D (2004) The ecosystem approach. Coherent actions for marine and coastal environments. A Report to the UK Government. English Nature, Peterborough, p 65

LePage BA (2011) Wetlands: a multidisciplinary perspective. In: LePage BA (ed) Wetlands—integrating multidisciplinary concepts. Springer, Dordrecht

Maltby E (1986) Waterlogged wealth: why waste the world's wet places. Earthscan, London, p 200

Maltby E (2009a) The changing wetland paradigm. In: Maltby E, Barker T (eds) The wetlands handbook. Blackwell, West Sussex, p 800

Maltby E (2009b) Functional assessment of wetlands: towards a valuation of ecosystem services. Woodhead Publishing, Oxford, p 712

Maltby E, Hogan DV, McInnes RJ (1996) Functional analysis of european wetland ecosystems—phase I (FAEWE): the functioning of river marginal wetland ecosystems—improving the science base for the development of procedures of functional analysis. Final Report EC DG XII CT90-0084. EUR 16132. Office of the Official Publications of European Communities, Brussels, Belgium, p 448

McInnes RJ (2007) Integrating ecosystem services within a 50-year vision for wetlands. Unpublished WWT Report to the England Wetland Vision Partnership, Slimbridge, p 34

McInnes RJ (2008) Why wetlands matter to people. Environ Sci 17:6–9

McInnes RJ, Maltby E, Neuber MS, Rostron CP (1998) Functional analysis: transforming expert knowledge into a practical management tool. In: McCabe AJ, Davis JA (eds) Wetlands for the future. Gleneagles Publishing, Adelaide, pp 407–429

Middleton B (2011) Multidisciplinary approaches to climate change questions. In: LePage BA (ed) Wetlands—integrating multidisciplinary concepts. Springer, Dordrecht

Millennium Ecosystem Assessment (2005) Ecosystems and human well-being: wetlands and water synthesis: a report of the Millennium Ecosystem Assessment. World Resources Institute, Washington, p 68

Minich N (2011) The role of landscape architects and wetlands. In: LePage BA (ed) Wetlands—integrating multidisciplinary concepts. Springer, Dordrecht

Mitlin D, Thompson J (1995) Participatory approaches in urban areas: strengthening civil society or reinforcing the status quo? Environ Urban 7:231–250

Morris J, Gowing DJG, Mills J, Dunderdale JAL (2000) Reconciling agricultural economic and environmental objectives: the case of recreating wetlands in the Fenland area of eastern England. Agric Ecosyst Environ 79:245–257

Nelson E, Mendoza G, Regetz J, Polasky S, Tallis H, Cameron DR, Chan KMA, Daily GC, Goldstein J, Kareiva PM, Lonsdorf E, Naidoo R, Ricketts TH, Shaw MR (2009) Modeling multiple ecosystem services, biodiversity conservation, commodity production and tradeoffs at landscape scales. Front Ecol Environ 7:4–11

Nyman JA (2011) Ecological functions of wetlands. In: LePage BA (ed) Wetlands—integrating multidisciplinary concepts. Springer, Dordrecht

Office of Technology Assessment (1984) Wetlands—their use and regulation, OTA-O-206. United States Government Printing Office, Washington, p 208

Olsson P, Folke C (2001) Local ecological knowledge and institutional dynamics for ecosystem management: a study of Lake Racken Watershed, Sweden. Ecosystems 4:85–104

Olsson P, Folke C, Berkes F (2004) Adaptive co-management for building resilience in social–ecological systems. Environ Manage 34:75–90

Pound D (2009) The eco what approach? ECOS 30:17–27

Richardson CJ, Hussain NA (2006) Restoring the Garden of Eden: an ecological assessment of the marshes of Iraq. BioScience 56:477–489

Richardson CJ, Reiss P, Hussain NA, Alwash AJ, Pool DJ (2005) The restoration potential of the Mesopotamian marshes of Iraq. Science 307:1307–1311

Robertson MM (2004) The neoliberalization of ecosystem services: wetland mitigation banking and problems in environmental governance. Geoforum 35:361–373

Salyer JC II, Gillet FG (1964) Federal refuges. In: Linduska JP (ed) Waterfowl tomorrow. United States Government Printing Office, Washington, pp 497–508

Scholz M, Harrington R, Carroll P, Mustafa A (2007) The Integrated Constructed Wetlands (ICW) concept. Wetlands 27:337–354

Sergio F, Pedrini P, Marchesi L (2003) Reconciling the dichotomy between single species and ecosystem conservation: black kites (*Milvus migrans*) and eutrophication in pre-Alpine lakes. Biol Conserv 110:101–111

Shepherd G (2004) The ecosystem approach: five steps to implementation. International Union for Conservation of Nature, Gland, Switzerland, p 30

Springate-Baginski O, Allen D, Darwall WRT (2009) An integrated wetland assessment toolkit: a guide to good practice. International Union for Conservation of Nature, Gland, Switzerland, p 144

Thesiger W (1964) The marsh arabs. Penguin Classics, London, p 256

Turner RK, van den Bergh JCJM, Söderqvist T, Barendregt A, van der Straaten J, Maltby E, van Ierland EC (2000) Ecological-economic analysis of wetlands: scientific integration for management and policy. Ecol Econ 35:7–23

Turner RK, van den Bergh JCJM, Brouwer R (2003) Introduction. In: Turner RK, van den Bergh JCJM, Brouwer R (eds) Managing wetlands: an ecological economics approach. Edward Elgar, Cheltenham, pp 1–16

United States Congress Public Law (1954) Watershed Protection and Flood Prevention Act. US Congress PL566, August 4, 1954, 68 Stat 666

Vira B, Adams WM (2009) Ecosystem services and conservation strategy: beware the silver bullet. Conserv Lett 2:158–162

Walker B, Carpenter SR, Anderies J, Abel N, Cumming GS, Janssen M, Lebel L, Norberg J, Peterson GD, Pritchard R (2002) Resilience management in social–ecological systems: a working hypothesis for a participatory approach. Conservation Ecology 6:14 [online]. www.consecol.org/vol6/iss1/art14

Wood A, van Halsema GE (2008) Scoping agriculture-wetlands interactions towards a sustainable multiple-response strategy. Food and Agriculture Organization of the United Nations, Water Report No. 33. Food and Agriculture Organization of the United Nations, Rome, p 155

Young G (1977) Return to the marshes: life with the marsh Arabs of Iraq. Collins, London, p 224

Chapter 13
The Role of Landscape Architects and Wetlands

Nancy A. Minich

Abstract The broad educational background, experience, and employment history of a landscape architect encompasses the elements of design and engineering, thereby bridging the traditional licensed professional fields of architecture, engineering, and natural sciences that are typically involved in the design and management of wetlands. The landscape architect's education and experience is typically diverse and provides a broad and somewhat unique multidisciplinary perspective to projects. On wetland creation, enhancement, and restoration projects they convey important design and aesthetic components to the regulatory agencies and general public. As environmental and regulatory issues come to the forefront of wetland design and planning, landscape architects are playing an increasingly greater and perhaps more important role in the design, restoration, and management of wetland projects. Understanding the science of wetlands as well as their aesthetic application in wetland construction and stormwater management will become more important, especially for those projects located in densely populated areas of major cities and suburbs. For the general public, graphic communication of the design is critical to acceptance of the project. With the advent of more realistic computer programs the reality of the future wetland landscape has become more apparent and therefore, visually understood by the client. In many instances, the success of a project, regardless of the total cost, hinges directly on aesthetics and how the societal and ecological benefits are communicated to, and understood by the public.

The Landscape Architect's Training and Expertise

As a profession landscape, architecture is relatively young in comparison to the allied professions of engineering and architecture. The emerging environmental field has presented landscape architects with the opportunity of leading or at least

N. A. Minich (✉)
NAM Planning and Design, LLC, 6575 Greenhill Road, Lumberville, PA 18933, USA
e-mail: nminich6575@comcast.net

B. A. LePage (ed.), *Wetlands,*
DOI 10.1007/978-94-007-0551-7_13, © Springer Science+Business Media B.V. 2011

playing a major role in complex wetland projects that involve inventory, analysis, design, regulatory review, and project management.

According to the American Society of Landscape Architects (ASLA), the national professional organization for landscape architects, "Landscape architecture is the profession that encompasses the analysis, planning, design, management and stewardship of the natural and built environment.... Landscape architects plan and design traditional places such as parks, residential developments, campuses, gardens, cemeteries, commercial centers, resorts, transportation facilities, corporate and institutional centers and waterfront developments. They also design and plan the restoration of natural places disturbed by humans such as wetlands, stream corridors, mined areas and forested land." (American Society of Landscape Architects 2010a).

Until recent years, landscape architects were trained more in the use of ornamental plants in the landscape, with relatively little study or understanding of the complexity of native plant communities or the natural systems that are part of a wetland site's ecology. Previously, they were involved in a wetland project usually at the end to meet a mandated development ordinance or to embellish a man-made feature. Currently, landscape architects are now perceived by their science, design, and construction peers as an important participant in the planning, design, and construction process, and on equal ground. They have become more than just "decorators" of the landscape and more involved in the initial stages of a design or restoration project, contributing more to the functional aspects of a site design. The regulatory agencies are more apt to understand a wetland project and thus approve them when a landscape architect is involved. Despite the use of sound science, complex wetland projects benefit considerably from the graphic presentations illustrating the implementation and use of Best Management Practice (BMPs) or system features and the landscape architect's explanation of their selection and use.

Environmental Sustainability

Environmental decline and natural resource degradation, particularly wetlands, has become common knowledge to most Americans. The public has become considerably more educated on environmental issues in recent years and subsequently more concerned about them. Although the 1970's opened the environmental field and increased the average American's awareness of the state of the environment, it was not until the twenty first Century that environmental issues were brought to a universal level of understanding and equated to societal values. As sustainability and "green" projects become a major focus in people's lives, the diverse skills of landscape architects should be used to communicate the importance of wetland functions and values to the public.

For the public to understand the benefits of wetlands and appreciate their intrinsic beauty, the landscape must have visual appeal in addition to good science to prevent an engineered look. The man-made, structural features of designed wetlands (including streams) such as pipes, baffles, and non-indigenous rock (rip-rap) that have traditionally been used to support the functions of natural wetlands now need to be designed to look more natural and less contrived. In the past, wetlands were perceived by the public as weedy or in more derogatory terms as mosquito-infested swamps without regard for their value as a habitat, flood control, or filtering pollutants. Like other allied professionals, landscape architects regarded wetlands as low-preference features in the landscape and often left them to be engineered with little appeal or interest to the public. Presently wetlands are viewed as high-priority landscape features and every effort is being made to preserve wetlands or to construct, enhance, or restore them to look as natural as possible while implementing approaches to engage and educate the public in their design and benefits. Educational materials such as signage or pamplets are being installed in community wetlands to highlight their importance and the ecological and economic benefits that wetlands bring to community. For example, the wetland sign installed at the New York Botanical Garden in Bronx, New York explains the importance of graphic communication to engage visitors in the wetland area (Fig. 13.1).

Landscape architects now are cognizant that they need to soften the engineered, man-made features with other non-structural natural features such as plants. Site

Fig. 13.1 A sign installed at the New York Botanical Garden in Bronx, New York that provides an explanation on the importance of wetlands that is written for public consumption

grading and topography is important so as to emulate natural contours and hydrology of the region. Other human interest features that could be included are pathways that direct and expose visitors to the most scenic views of the wetland, while avoiding or concealing the unattractive, yet essential features like an overflow drain. The landscape architect's skills are essential in concealing these elements with natural elements such as low-growing ferns and native grasses and wildflowers or designing other features to make such essential, but unappealing features less visible.

It is the landscape architects' well-rounded background and ability to effectively communicate with the public that makes them poised for taking a major role in such projects. The US News and World Report magazine listed landscape architecture as one of the best careers in 2009. Subsequently, this profession will continue to flourish and the US News and World Report (2009) stated "Governments and nonprofit groups are restoring increasing amounts of land to their primitive states. This trend will very likely accelerate in the new administration and Congress, and with environmentalism ever growing."

Landscape Architect's Licensure

Landscape architects are similar to professional engineers and architects in that they are required by individual states to be licensed to protect the public and establish a basic level of competency. In the United States, as of January 2011 there are licensure laws regarding licensed landscape architects to practice and, or to use the title "landscape architect" in all 50 states. The definition of practice and title varies from state to state. The definition is in each state's board of licensure for regulating and licensing landscape architects. The purpose of the licensure is to protect the health, safety, and welfare of the public. Each state has a licensure board that defines the minimum requirements for the practice of landscape architecture. This is generally based on established minimal requirements of receiving a degree from an accredited landscape architecture program, several years of supervision or training under another landscape architect or related professional, and passing a rigorous national exam in all the areas of competencies in the profession. Once these criteria have been met, an individual becomes a registered or certified landscape architect depending on the state licensure law, and can seal and sign their drawings and specifications (American Society of Landscape Architects 2010b).

Most federal, state, and local municipalities require drawings and related professional work to be sealed or stamped and signed by the professional who performed the work on the drawings or in the related document. For instance, in the Commonwealth of Pennsylvania, the Bureau of Professional and Occupational Affairs, Department of State regulates licensure. However, once an individual becomes licenced, the individual is required to maintain their credentials by working in the field and remaining current with industry practices and new developments by enrolling in educational courses and workshops. These boards determine the number of continuing education credits that an individual must acquire every two years to renew the license as well as to govern practice and the conduct of the professional (Pennsylvania Department of State 2010).

Landscape Architect's Role in Stormwater Management

Most practicing landscape architects concur that the need for a landscape architect is determined by the goals and objectives of a client's project and the regulations within that state or local municipality. The landscape architect needs to be a key professional in developing the concept with an emphasis on aesthetics and human engagement in the development of a wetland landscape. Accordingly, the landscape architect will continue to serve a primary role in communicating the non-structural concepts (i.e., the use of native plants as a tool in wetland design and overall stormwater management) to their team peers, regulatory agencies, and the public. Past stormwater designs generally were highly structural elements that were not only expensive to install and maintain, but unappealing with their use of bare concrete, gabions, and exposed pipes. However, when plants began to become incorporated into designs, it was commonly with unnatural patterns and non-native species that limited their future success and attractiveness of the project. The movement towards the use of native plant species and simple, more natural designs is beginning to take hold in the industry.

Many clients now recognize the need for a professional landscape architect for the design of stormwater systems that conform to the ever changing federal, state, and local regulations. In these environmentally conscious times, landscape architects are well trained to address and implement the current and ever-changing stormwater regulations associated with the federal Clean Water Act into their designs. For example, in the 1970s when stormwater issues came to light, detention basins were constructed throughout the country that addressed the quantity of stormwater being detained (Fig. 13.2). Water quality was not ever considered as being important at that time.

Today, water quality is now at the forefront of stormwater management and basins such as those constructed in the 1970s are being designed or re-designed to infiltrate water and with native plants to remediate nutrients and contaminants and reduce maintenance. Bioswales and vegetated buffer strips are now being incorporated into industrial, commercial, and public stormwater basin designs. Figure 13.3 provides an illustration of a bioswale that was designed for a parking lot in New Hope, Pennsylvania. This design illustrates the use of non-structural created wetlands as a cost-effective approach to treat stormwater runoff. In this case the parking lot was graded towards the bioswale islands to store and treat the nonpoint source (NPS) pollutants derived from the vehicles that use the parking lot. The native plants perform uptake of the NPS pollutants and create an attractive series of garden-like landscapes. The engineering of the bioswale is not apparent to the average person as the native plants are functional and aesthetically pleasing. The overflow structures are well hidden by the plant community for most of the year and the costs

Fig. 13.2 An example of a typical stormwater detention basin. These types of stormwater basins were built to address the quantity of stormwater being detained rather than improving water quality. Note the concrete headwall and rip-rap apron that is typical of 1970's style detention basins. Current practices such as mowing the lawn to the water's edge contribute to poor water quality as is evident by the abundant patches of green algae on the water surface

associated with managing a traditional detention basin such as regular mowing and trash removal are eliminated. In addition, parking lots, which are generally not that aesthetically pleasing, are now an amenity enhanced by the native plant community that has many seasons of interest and color.

As wetlands and water quality became degraded over time, allied professionals realized that structural solutions were not only cost ineffective, they were not as successful as non-structural interventions and incongruous with the natural environment. The non-structural methods, such as wetland construction to treat stormwater prior to discharging into a creek, are less expensive and better in capturing the NPS pollution in stormwater through the use of vegetation to treat stormwater runoff. It is a widely accepted fact that specific wetland plants can successfully treat the most difficult part of stormwater to treat the NPS pollution that is a result of runoff largely from agricultural fields, suburban lawns, and parks and recreation areas that use fertilizers, pesticides, and herbicides as well as the chemicals from roads and parking lots. Plant communities are a natural system for filtering pollutants especially those entrained in stormwater runoff. Understanding these plant communities as well as their aesthetic applications in wetland construction and stormwater management will become a growing requirement by the regulatory agencies, especially those located in densely populated areas of major cities and suburbs.

Landscape architects will continue to play a significant role in designing natural systems that provide better contact time to capture and treat the NPS pollution in constructed wetlands and related wetland systems such as retention basins and vegetated wetland systems. Terri Bentley, a planner who was a member of the advisory committee for the development of the *Pennsylvania Best Management Practices*

Fig. 13.3 An example of a bioswale integrating non-structural BMPs that was recently constructed in a public parking lot in New Hope, Pennsylvania. These types of BMPs infiltrate stormwater back into the ground rather than into a municipal stormwater system. **a** Conceptual plan view. **b** Bioswale showing curb cut where water from the parking lot drains. **c** Incorporation of bioswale with the ornamental structural elements of the parking lot design. **d** Newly planted bioswale. **e** Bioswale after 8 years of growth

Manual (Department of Environmental Protection, Bureau of Watershed Management 2006) believes landscape architects can play a major role in the design of wetlands. They are one of the few professionals in the wetland field who can design for specific user needs and tailor the plant list to complement the site conditions (1 April 2009, Personal Communication). The federal National Pollution Discharge Elimination System (NPDES) regulations require onsite management of the quality and quantity of stormwater, thereby encouraging and inspiring landscape architects to design artful ways to create functional wetlands as part of the site design. As a result, landscape architects have been at the forefront of designing such stormwater management systems for their clients.

Furthermore, many state boards of professional licensing are granting landscape architects the authority to seal drawings for specific stormwater management designs and plans that are based on the approved BMPs. Subsequently, the role of the landscape architect is becoming more entrained in wetland design and essential to the planning team. Up until recently, only professional engineers had the legal authority to design and implement many of the structural and non-structural BMPs and to sign and seal certain types of stormwater management designs.

Landscape architects are also able to enhance the level of the Leadership in Energy and Environmental Design (LEED) certification of a site by designing environmentally sustainable construction projects such as a constructed wetland instead of a traditional detention basin to filter NPS pollution. Landscape architects are among the select professionals eligible to become LEED certified, thus making them a more significant professional team member.

Designing for Human Needs and Acceptance

As most professionals recognize, wetland design and especially creation, is not the purview of any one profession. Each wetland-related professional brings to a project a particular expertise that is essential to the aesthetic and functional aspects of a successful wetland project. Although engineers and scientists have the scientific acumen to calculate and develop the functional aspects of wetlands, the landscape architect will partake in the collaboration of all aesthetic and functional components of the design. When changes are made during the course of the wetland project design by one of the technical team members, the landscape architect can oversee and coordinate the changes and modify the design accordingly. This follows the model outlined in LePage (2011). The landscape architect can interpret the technical information to the client through animated sketches and colorful rendered drawings.

The landscape architect's role overlaps with the natural sciences and other technical professionals as well as engineering. However, the landscape architect focus is more on how to design a wetland that has a high level of aesthetics and thus becomes a preferred landscape feature or focal point where there was once a degraded landscape. More importantly, landscape architects are well positioned to understand and effectively manage the technical jargon of numerous the professional disciplines involved in the project and communicate the issues that arise in a manner that is understandable to the project team in a non-adversarial manner.

More importantly though, the success of a project wetland project, especially those located in urban settings where the public is intimately involved is contingent on the public's understanding of the project's benefits. In many cases, the wetland-related professionals possess a limited skill set in training and human experience. Nor do they realize the importance of a wetland as an aesthetic experience. As pointed out by Kaplan (2011) effective communication is essential to the success of any project. Landscape architects are well trained by way of both education and experience to develop an engaging presentation to the client and public that solves the basic communication problems and also appeals to their complex human needs. That is, the importance of nature to the human consumers of a wetland project.

Landscape architects spend considerable time studying established research and sociology on people's needs and interests. Recent environmental psychology research has concluded that people have preferences for the natural environment. Rachel and Stephen Kaplan have identified people's preferences related to water. The Kaplans' research indicates that water is an attractive landscape feature, and how the water feature is designed will impact how people perceive the water and ultimately the project in which the wetland is sited (Kaplan et al. 1998). Furthermore, Joan Nassauer, a landscape architect, has conducted research in determining how to design native landscapes that will be more accepted and appreciated by the general public more so in the suburban landscape (Nassauer 1995). Natural area restorations are sometimes criticized as looking too unkempt when placed in an urban context, and elements of design are often suggested to provide the public with visual cues to signal that these sites are in fact being cared for (Nassauer 1995).

There tends to be inherent merit in the use of the non-native, highly manicured landscape stewardship model, albeit it is non-sustainable and antiquated in today's environmentally conscious times. Without indications that it is not abandoned or cared for, the native wetland paradigm has difficulty standing on its own merits in most neighborhoods. This in part is due to the public's long-held negative perception of wetlands. Examples abound in urban and suburban areas where a restored wetland has been retrofitted into an otherwise acceptable model landscape with large areas of mowed lawns and symmetrical trees in the non-ecological style of an English nineteenth Century park. Conversely, the more sustainable wetland system has been viewed as a bad or an unhealthy and unattractive place to avoid because of the "weedy" native vegetation and the perception that wetlands breed mosquitos and other noxious insects. If the wetland is functioning properly, then the ecosystem should not contain offensive odors and mosquitos. Nassauer (1995) recommends that landscape architects design 'cues to care' such as a neatly mowed edges adjacent to native wetlands, signage, fence posts, gateways, or bridges to let the public know that this landscape, especially the wetland, is native and subsequently maintained. These recognizable features assuage the average person's need to want more order or to over manicure and weed. Landscape architects are urged to place attractive educational signage at restored or created wetlands to inform the public on the purpose of the wetland and the ecological and economic benefits it will bestow to the community. Without education, the average person will continue to see the wetland as low-preference green space if they do not understand its purpose and value in an otherwise degraded hydrological system of checks and balances.

Such is the case of a brownfield site located outside of Philadelphia (Fig. 13.4 [before]). The municipality that owned the contaminated site wanted to have it restored to a park-like setting. The design issues that needed to be resolved as part of this project included the removal of historic contaminants, retoration of a well-utilized neighborhood park, a flashy degraded urban creek, and lack of stormwater storage along the banks of this creek because of the filling in of the floodplain. As part of the public outreach program a landscape architect was retained to meet the municipality's objectives and provide design ideas and graphic renditions of the completed project. The municipality had two criteria for this project: (1) to ensure that the park was designed to meet the surrounding community needs, and (2) to ensure the design was in accordance with the management options of a rivers conservation plan that is on the Pennsylvania Rivers Registry. NAM Planning & Design, LLC and Symbiosis, Inc. collaborated on this brownfield park design that was consistent with the goals of the Tookany Creek Watershed Management Plan

Fig. 13.4 A brownfield site located in the Philadelphia suburbs. The site was being used as a municipal park. Tookany Creek is a degraded and flashy urban stream that is located to the *right* of the image. Stratigraphic data indicated that up to 15 ft. of fill was placed in the floodplain to provide more useable area in the park. As a result, the floodplain was lost and the streambank consisted of a vertical stone wall that was approximately 10 ft. high

and linked this portion of the creek to the Township's trail system. Aligned with mitigation plan the landscape architect team designed a sustainable park featuring benches, lighting, meadows, and shallow rain gardens that contained stormwater on-site, while providing an educational wetland inlet to accommodate excessive flows and treat NPS pollution in the creek.

The site was re-contoured to be more visually interesting from the engineer's balance of the cut and fill (Fig. 13.5). Several scenarios of design grading were

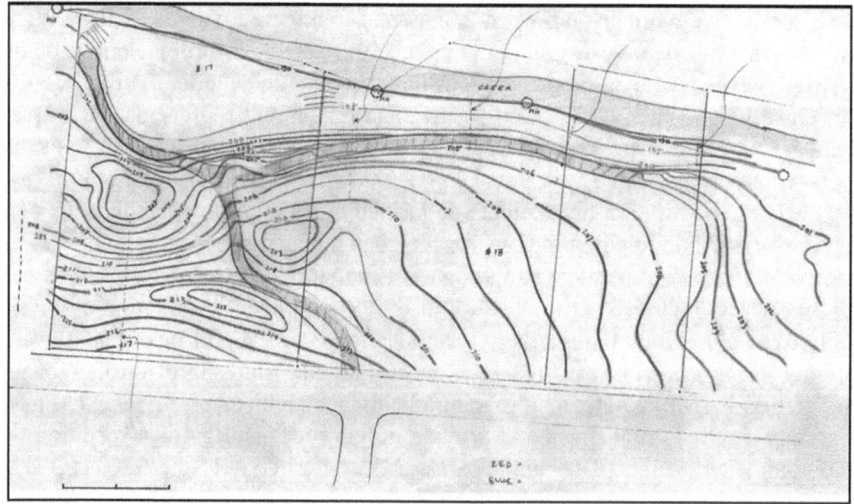

Fig. 13.5 Conceptual sketch of the brownfield site showing a preliminary grading plan based on anticipated amount of soil that would be excavated and removed in specific certain areas versus the areas that would require the addition of soil to meet the topographies of the areas adjacent to the project site

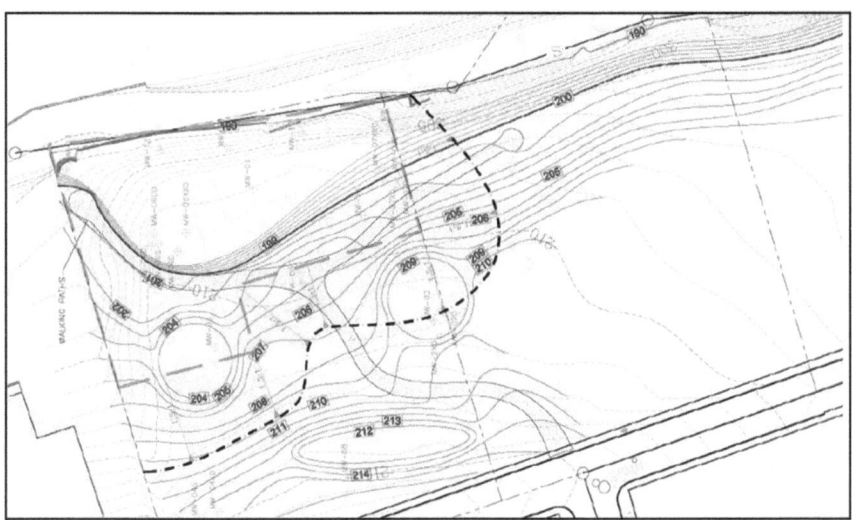

Fig. 13.6 A more refined conceptual sketch that incorporates the aesthetic and environmental elements with the required engineering elements such as surface topographies and the volume and velocity of the water flowing in the stream following rain events. The *dotted line* delineates the area of excavation

developed to incorporate the environmental requirements into the final design (Fig. 13.6). Other significant human ecology features beyond the regulatory issues were the path system that linked to the adjacent neighborhood for dog walkers and to the nearby train station. A series of overlooks along the creek highlighted the importance of the site remediation and its value to the larger watershed. The landscape architects goal was to restore the site to its historic topography. Educational signs were proposed to explain the functions of wetlands and floodplains and how the restoration returned their original functions. The paths meandered around small retention ponds that were planted with native plants to capture rainwater and a wildflower meadow to treat the NPS pollution runoff from the frequently flooded streets. One of the more important roles in this project was to develop an appealing watercolor rendering of the proposed design (Fig. 13.7). The landscape architects led the discussion with the municipality highlighting the community and natural resource benefits of the park design. The graphic perspective rendering was a major communication tool for the ideas of the plan.

The design and presentation of these characteristics is the domain of landscape architects. They can communicate the aesthetics of a site via skillfully executed and colorfully rendered drawings. Landscape architects spend the majority of their time in landscape architecture programs learning how to best communicate the abstractness of a design to appeal to the public and have always excelled in graphic communication as part of the client presentation. Hand drawn or computer-generated renderings are typical means of graphically communicating design ideas. Most engi-

Fig. 13.7 A *watercolor* rendering of the final design showing a plan, landscape, and cross-sectional views

neers, architects, landscape architects, and other professionals work predominantly in plan view, which is often confusing to the average client. Perspective drawings of key locations clearly communicate the design intent and can show special features unable to be understood from the plan view. Perspective drawings of key locations bring the design to life; a picture is truly worth a 1,000 words.

Most hand drawn renderings require extensive time to accurately set up the perspective that involves accurate details drawn in color. Once completed, hand drawings cannot be redone easily, unlike the computer-generated renderings. Computer-generated perspectives can be very accurately developed in perspective from a scaled plan. Additionally, if any changes are necessary in the next round of presentations, it takes a fraction of time to alter the drawing accordingly.

With the advent of more realistic computer programs such as SketchUp, Vectorworks, Rhino, and Photoshop, the reality of the future wetland landscape has become more apparent and therefore, visually understood by the client. Because of the three-dimensional nature of these programs, actual photos and real people can be used in the computer rendering to animate the proposed landscape and personalize the project for the client.

References

American Society of Landscape Architects (2010a) http://www.ASLA.org

American Society of Landscape Architects (2010b) http://archives.asla.org/members/govtaffairs/licensure/licensure_toc2.html

Department of Environmental Protection, Bureau of Watershed Management (2006) Pennsylvania stormwater best management practices manual. Technical Guidance 363-0300-02. Pennsylvania Department of Environmental Protection, Bureau of Water Management, Harrisburg, p 685

Kaplan R (2011) Acknowledging and benefitting from multiple perspectives. In: LePage BA (ed) Wetlands—integrating multidisciplinary concepts. Springer, Dordrecht

Kaplan R, Kaplan S, Ryan RL (1998) With people in mind. Design and management of everyday nature. Island Press, Washington, p 239

LePage BA (2011) Wetlands: a multidisciplinary perspective. In: LePage BA (ed) Wetlands—integrating multidisciplinary concepts. Springer, Dordrecht

Nassauer JI (1995) Messy ecosystem, orderly frame. Landsc J 14:161–170

Pennsylvania Department of State (2010) www.dos.state.pa.us/bpoa/cwp/view.asp?A=1104&Q=432638

US News and World Report (2009) http://www.usnews.com/articles/business/best-careers/2008/12/11/best-careers-2009-landscape-architect.html

Chapter 14
Floating Islands—An Alternative to Urban Wetlands

Lanshing Hwang and Ben A. LePage

Abstract The creation, enhancement, and restoration of wetlands in the urban set-ting pose a suite of challenges including space constraints, sediment contamination, upstream pollution sources, and compromised water quality for wetland practitio-ners. These challenges are amplified in situations where the wetlands are dominated by fluvial systems (e.g., streams, rivers). Here we discuss the installation of seven floating islands in a highly impacted (polluted) part of the tidal Anacostia River, adjacent to Diamond Teague Park, Washington, D.C. Floating islands (also called floating wetlands) are based on natural phenomena and have been used to improve water quality in many parts of Asia. However, they are only just being consid-ered as viable wetland creation options in the United States. Although the surface area of the seven floating islands is small, measuring slightly over 1,600 square feet, the roots floating in the water column provide an equivalent wetland area of six acres. The nitrogen, phosphate, and ammonia that these floating islands have been designed to remove from the Anacostia River annually are 990, 138, and 990 pounds, respectively. In addition to improving water quality, these wetlands create valuable wildlife and aquatic habitat. Floating islands provide an innovative oppor-tunity in urban areas where issues such as water quality, site constraints, aesthetics, community needs, and costs need to be considered.

Continued development in the ever shrinking urban setting has challenged the skill and innovation of the landscape architect to work with sites that are in many ways compromised. Many urban brownfields are often best described as blights on the landscape and generally characterized by poor water and soil qualities. In some cases the soil and water conditions in brownfield development projects verge on being hazardous. While there are various approaches to address and mitigate such sites located in upland settings, lowlands such as wetlands, streams, and waterfronts

L. Hwang (✉)
Symbiosis Inc., 9008 Brae Brook Drive, Lanham, MD 20706, USA
e-mail: lanshingh@symbiosis-la.com

B. A. LePage (ed.), *Wetlands,*
DOI 10.1007/978-94-007-0551-7_14, © Springer Science+Business Media B.V. 2011

Fig. 14.1 Examples of floating islands. (Courtesy and with permission of BioHaven® International)

come with a special set of challenges, and remediation and mitigation options to improve the environmental conditions, create habitat, improve the water quality, and develop a site that is a esthetically pleasing are limited.

Improvement of water conditions is particularly challenging given that impacts to water quality and aesthetics may be due to activities occurring upstream of a project site. Regardless of blame, improving water quality should be everyone's responsibility, no matter how trivial the approach. A recent innovation aimed at improving water quality is the creation of floating islands (also called floating wetlands). Floating islands are masses of floating aquatic plants that are a few inches to several feet thick (Fig. 14.1). Essentially they function no differently from wetlands, except for the fact that these emergent wetlands are floating and the plant roots hang into the water column. The plants remove contaminants from the water through filtration and uptake of nutrients and contaminants into the plant's tissues (Fig. 14.2). Biofilms are composed predominantly of bacteria (aerobic and anaerobic) and are present on the surface of the plant's roots. These films aid in the removal of ammonium, nitrate, phosphate, and heavy metals (Stewart et al. 2008). While floating islands are relatively common natural phenomena that are found in marshlands, lakes, and similar wetland environments throughout the world, the implementation of man-made floating islands for improving water quality is only just beginning to be considered and implemented as a viable water treatment solution (Kadlec and Wallace 2009).

The Diamond Teague Park discussed in this chapter is one example of putting floating islands into practice to help improve the aesthetics and water quality along a significantly impacted tidal portion of the Anacostia River in Washington, DC. Although the park is located on the only piece of natural shoreline in the area that is not bulk-headed, it regrettably has a long history of industrial use and environmental issues. The park is situated adjacent to the District of Columbia Water and Sewer Authority (WASA), and properties located upstream and downstream of the park carry petroleum, heavy metals, and other pollutants into the river with each stormwater event. In addition, there are several major combined sewer outfalls (CSOs) discharging into the river within the boundary of the park. Water quality is poor following runoff events and the smell is foul during the summer months. Immediately offshore, the United States Environmental Protection Agency (USEPA)

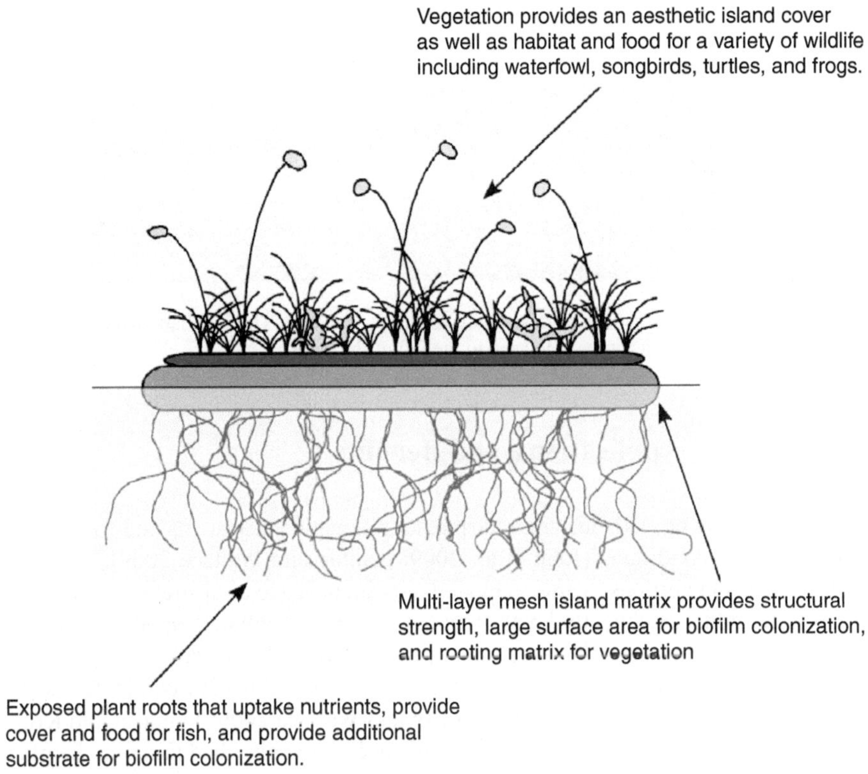

Vegetation provides an aesthetic island cover
as well as habitat and food for a variety of wildlife
including waterfowl, songbirds, turtles, and frogs.

Multi-layer mesh island matrix provides structural
strength, large surface area for biofilm colonization,
and rooting matrix for vegetation

Exposed plant roots that uptake nutrients, provide
cover and food for fish, and provide additional
substrate for biofilm colonization.

Fig. 14.2 Illustration showing the three major components of floating islands; the vegetation, mesh matrix, and plant roots

has implemented several pilot sub-aqueous capping projects to test capping materials for polluted sediments on the river bottom. Although submerged aquatic vegetation (SAV) was once plentiful along this section of the river, since the turn of the last century pollution and poor water quality has virtually eliminated SAV communities from the tidal portion of the Anacostia River (United Marine International 2010).

Although the project site is only a little over an acre in size, it is situated in the core of the City's new economic development focus area. The City envisioned it as a waterfront gateway to the revitalized southeast waterfront of Washington, DC. Phase I of the program includes a water taxi dock to transport baseball fans to the new baseball stadium, an environmental pier for waterfront education, a boardwalk that is a portion of the future Anacostia Riverwalk, and most of all an urban waterfront park for the citizens of Washington (Fig. 14.3). The park is named after Diamond Teague, a former member of Earth Conservation Corps and will have a memorial to honor his volunteer work that is closely associated with the natural history and ecology of the Anacostia River.

Fig. 14.3 Photograph of the area designated as the project site. The baseball stadium is in the background and the project area is located to the left of the wharf

Background, Applications, and Benefits

Floating islands or wetlands are widespread in aquatic ecosystems and form under most climatic conditions (John et al. 2009; Kadlec and Wallace 2009). They develop when: (1) portions of emergent wetlands anchored to a substrate grow out into deep water and subsequently break free; (2) portions of submerged aquatic vegetation become delaminated from the substrate and float to the surface; or (3) pieces of aquatic plants become a nucleus, attracting other vegetation and floating debris that ultimately grows to form an island (Kadlec and Wallace 2009; Vymazal and Kröpfelová 2008). The lack of soil/sediment and specialized air-filled tissue in the plant roots called aerenchyma assist in providing buoyancy for the islands.

While the use of constructed floating islands has been primarily focused on engineered waste- and stormwater treatment systems, there is growing interest in utilizing these artificial ecosystems for improving water quality where it has been compromised and intensive landuse precludes constructing traditional wetlands (DeBusk et al. 2005; Headley and Tanner 2006; Nahlik and Mitsch 2006; Stewart et al. 2008; Li et al. 2009; Hubbard 2010). Constructed floating islands such as BioHaven® Floating Islands developed by *Floating Island International* (www.floatingislandes.com) use the same principles as naturally-occurring floating islands (Figs. 14.2 and 14.4). The floating matrix is composed of non-woven plastic layers made from recycled polyethylene terephthalate (PET) bottles that area held together with inert marine foam. The mat provides floatation and a suitable matrix for supporting the plants. The vegetation planted on each mat can be tailored to meet the specific biological conditions of the water body and desired performance criteria such as nutrient and/or contaminant removal.

The environmental benefits of floating islands are significant. *Floating Island International* (www.floatingislandes.com) estimates that a square foot of 8-inch thick BioHaven® Floating Island provides 198 sq ft. of surface area and a 266 sq ft. island is equivalent to 1 acre of surface area. The increased surface area is the key to the effectiveness of floating islands. Within days of being deployed, all of the plant root and matrix surfaces are covered with biofilm. The algae, bacteria, fungi,

Fig. 14.4 Closeup of the floating mat with the plant roots protruding into the water. (Courtesy and with permission of BioHaven® International)

and other microbes that comprise the biofilm feed on the nutrients and contaminants that are dissolved or suspended in the water column. Similarly, the plant roots take up dissolved nutrients such as phosphorus and nitrogen, converting them into biologically useful forms (Mitsch and Gosselink 1986), while symbiotic fungi in the plant roots called mycorrhizae sequester heavy metals that would otherwise be harmful to the plants and bind them to the fungal cell walls (Galli et al. 1994; Göhre and Paszkowski 2006). The process of removing contaminants in soil, sediment, or water using plants is called phytoremediation.

In addition to the benefits of remediation, floating islands provide much needed wildlife and aquatic habitat. The complexity of the non-aquatic food webs established on floating islands is largely determined by the size of the islands and the level of vertical structure created through plant diversity. Simple food webs may only support small populations of insects and transient insect-eating birds, while more complex food webs may be able to support resident bird and small mammal populations. The complexity of the aquatic food webs is largely a function of the size of the floating island, the amount of vertical structure in the plant community growing above the water, the density of the plant roots, and water quality. At the minimum, one would expect to find herbivores such as small crustaceans grazing on the biofilm, predaceous beetles feeding on the herbivores, and small fish feeding on the insects and seeking shelter in within the plant roots. In addition to the biological benefits, the islands provide much needed shade that helps cool the water.

Floating Island Concept, Design, and Implementation

Given the site's physical constraints, as well as the environmental issues associated with creating public space on a contaminated site, the prospect of integrating floating islands into the park concept and design provided a novel alternative for dealing with the issues of water quality, aesthetics, and habitat creation. That is, the floating islands would provide much needed emergent wetland habitat without utilizing the limited shoreline and transform the otherwise impacted shoreline into a enjoyable civic space, while the wetland plants would help improve water quality by sequestering contaminants and excess nutrients in the water. As pointed out by Kadlec and Wallace (2009), additional benefits of using floating islands include the removal of nutrients and contaminants from the water column, but not the substrate. Remobilization of contaminants buried in river sediments is always a major concern when sediment is disturbed. Although the floating islands do not solve the issue of sediment remediation, they do not contribute to contaminant remobilization. Secondly, contaminants from the water column that are incorporated in the plant tissues can be collected and properly disposed by harvesting the plant biomass at the end of each growing season. A conceptual drawing developed as part of the City's Phase 1 economic development focus area is provided in Fig. 14.5.

Meetings aimed at developing an innovative concept for this waterfront development project followed by rigorous discussions with the project team consisting of marine engineers, floating island designers, and landscape architects occurred throughout the design process. Although floating islands have been installed in ponds and small lakes they have never been installed in a tidal ecosystem, which posed a unique set of physical conditions and challenges that needed to be considered prior to agreeing on a design. To ensure that the floating islands were not transported upstream or downstream with the tidal flux the anchoring system needed to be tailored to the conditions of this water body. An extensive tidal and wave analysis was performed by Moffatt and Nichol, an internationally recognized marine engineering firm prior to development of the anchoring system. Once the analysis was completed the appropriate size, shape, quantity, locations of the floating islands on the river, and anchoring system was finalized in light of the desired design effect for the park. Once completed the project team was able to develop a pre-construction plan and specifications for the floating islands (Figs. 14.6 and 14.7). The planting plan and selection of plants was determined by the landscape architect and floating island designer. The plant list and design was developed by the renowned landscape gardeners Wolfgan Oehme and Carol Oppenheimer. The plants were selected for their ability to withstand the extreme tidal conditions and most of all, their beauty for the consideration of planting composition for this public park. The plants were arranged and planted in large masses to provide a bold effect and include: billowy sweet flags (*Acorus calamus*), sedges (*Carex* spp.), and rushes (*Juncus* spp.); large-leaved wetland plants such as pickerelweed (*Pontederia cordata*), duck potato (*Sagittaria latofolia*), and Canna lily (*Canna* spp.); as well as eye-catching flowering natives such as spatterdock (*Nuphar luteum*), Hibiscus (*Hibiscus* spp.) and *Iris* (*Iris* spp.).

BlueWing Environmental Solutions & Technologies, the regional representative of BioHaven® International in Maryland were contracted to supply and construct the floating islands. Construction of the floating island took place in July 2009 and was completed in two weeks. As with any water encroachment on a federally-regulated water body, a wetland permit issued by the United States Army Corp of Engineers and District of Columbia was required to install floating islands into the Anacostia

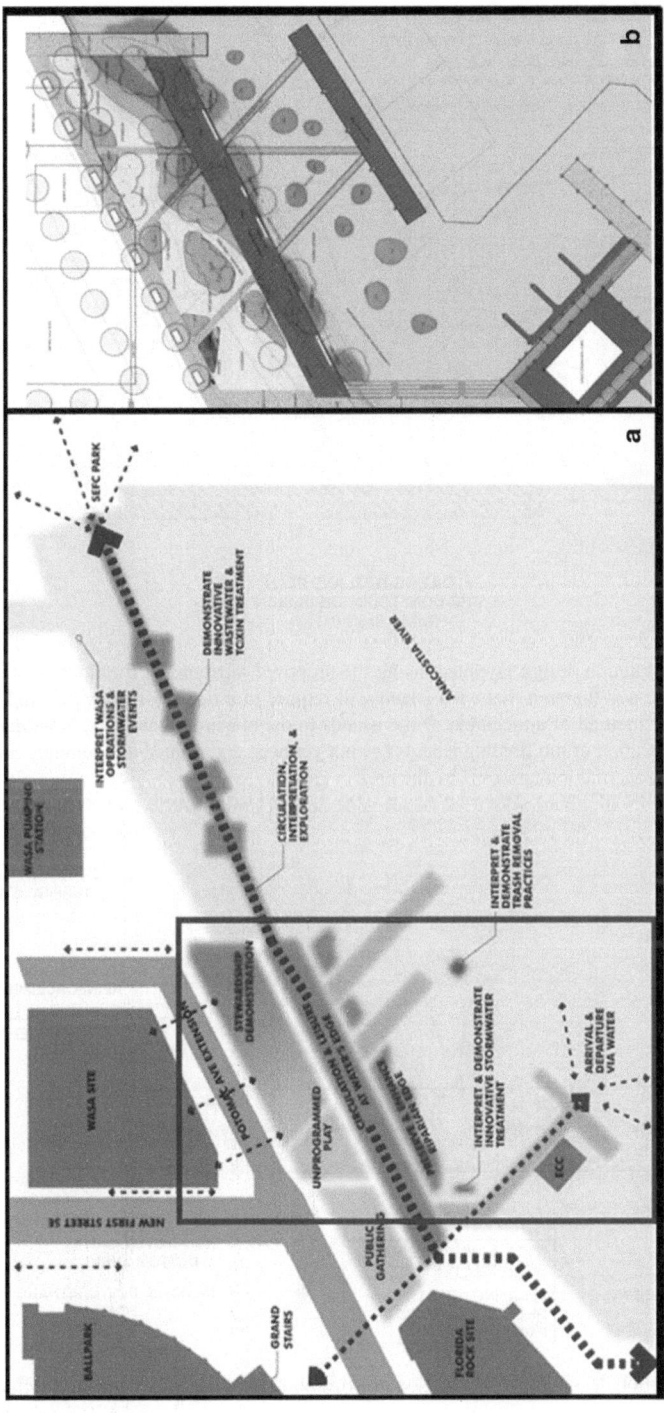

Fig. 14.5 a Conceptual drawing of Diamond Teague Park located on the shore of the Anacostia River. **b** Detailed view showing the shoreline, pier, and floating islands. When complete the park will include upland and water connections to the neighboring developments using a floating boardwalk. The floating boardwalk is necessary to offer continuity to the Riverwalk since there is no upland connection available due to the adjacent WASA operations. (Images are courtesy of the Landscape Architecture Bureau, Washington, DC)

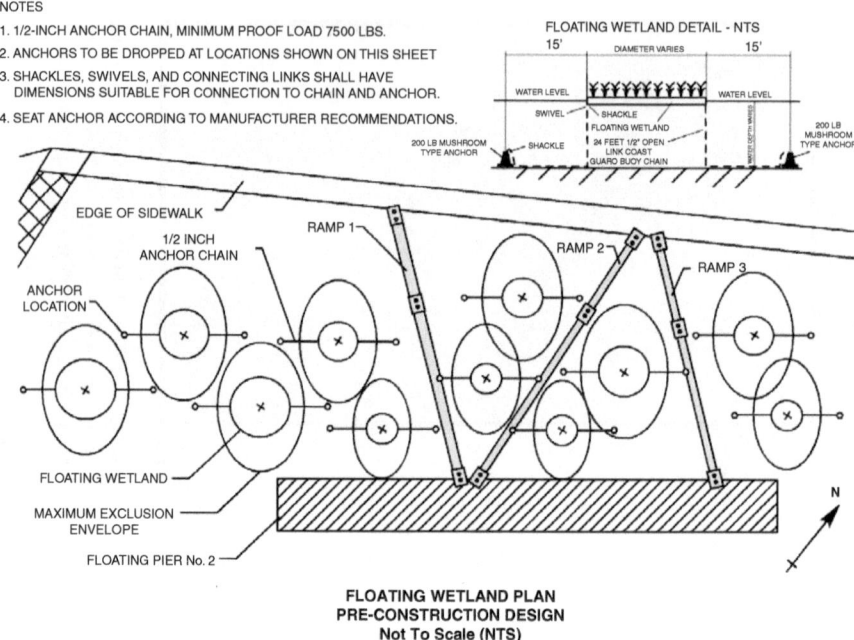

Fig. 14.6 Pre-construction design layout showing the proposed locations of the floating islands, access ramps, piers, and the area that each island will require in a tidal system (*top right*). The detail illustrates the method of attachment of the islands to the mushroom anchors. Note that the final number and location of the floating islands has not yet been determined and remains subject to final design and construction approval by the project team

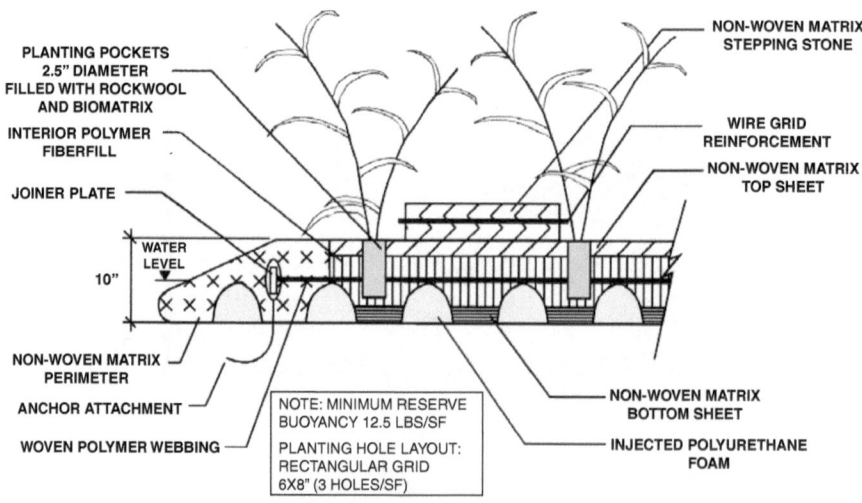

Fig. 14.7 Cross section design and specification drawing showing the components of the floating island

Fig. 14.8 The floating island modules awaiting assembly following delivery to the project site

River. Materials were delivered by truck and staged next to the pier (Fig. 14.8). The modules were then assembled on the pier (Figs. 14.9, 14.10, and 14.11). Seven circular islands ranging from 12 to 24 ft. in diameter were constructed. The modules were joined with pins, and swivels and shackles were installed at the anchor points. Finally, Biomix (a planting medium) was spread over the surface of the islands. On the larger islands stepping pads were installed to facilitate access for plant maintenance staff. Lightweight foam was injected at the locations where the stepping pads

Fig. 14.9 The floating island modules in the process of being assembled on the pier

Fig. 14.10 Assembled floating islands on the pier awaiting installation of the biomix and plants

Fig. 14.11 An assembled floating island that is ready for deployment showing the biomix, plants, and bird fencing

were installed and to increase buoyancy. The small diameter islands were planted with plant plugs while staged on the pier. The larger islands were assembled and launched; installation of the plants followed deployment (Fig. 14.12). The completed islands were then placed at regular intervals along the shoreline and attached to chains that were anchored to the river bottom with 200-lb. mushroom anchors

Fig. 14.12 A worker installing the biomix and stepping pads on a large floating island that was deployed. Note the *square* stepping pads and planting pockets (*circular holes*)

(Figs. 14.13, 14.14, 14.15, and 14.16). To prevent waterfowl from eating the vegetation, fencing was placed around each island (Figs. 14.11 and 14.13).

The 1,621 sq ft. of surface area provided by the seven floating islands provide a surface area that is equivalent to approximately 6-acres of wetlands, which far exceeds the surface area that was available for development along the shoreline. The predicted quantities of nitrogen, ammonia, and phosphorous to be removed annually are provided in Table 14.1.

Fig. 14.13 An assembled floating island showing the plants and bird fencing

Fig. 14.14 Completed floating island in the Anacostia River with the boardwalk, streambank plantings, and the ramp to the environmental pier in the background

Fig. 14.15 Completed floating islands in the Anacostia River with the boardwalk, water taxi dock access ramp, and the baseball stadium in the background

Fig. 14.16 Final placement of the floating islands relative to the pier and shoreline

Table 14.1 Floating island nutrient uptake analysis[a], Diamond Teague Park, Washington, DC

Island quantity	Island diameter (feet)	Island area (square feet)	Cumulative area (square feet)	Annual nitrogen removal (lbs.)	Annual phosphate removal (lbs.)	Annual ammonia removal (lbs.)
3	12	113.10	339.3	207.26	28.95	207.26
1	16	201.06	201.06	122.82	17.15	122.82
2	20	314.16	628.32	383.82	53.6	383.82
1	24	452.39	452.39	276.35	38.59	276.35
Total 7			1,621.06	990.25	138.3	990.25

[a] Analysis and data provided by BlueWing Environmental Solutions and Technologies

The maintenance and monitoring program is rather simple and consists of frequent inspections to ensure the structural integrity of the islands and plant health. No chemicals or fertilizers are used and regular weeding, especially during the initial months following deployment will allow the wetland plants to become well established. In the late fall, the plants will be harvested and properly disposed to prevent sequestered contaminants from re-entering the river.

Conclusions

The use of floating islands as an alternative to creating wetlands in an urban setting is a relatively new approach that is generating considerable interest, especially in the United States. During the course of this project, representatives of the Department of Environment of the District of Columbia visited the Park with an interest on potential applications for consideration of other projects in the City. As interest in utilizing this technology for improving water quality and habitat creation on impacted water bodies grows, it will be incumbent on the parties that employ floating islands to document and communicate projects that were successful, as well as those that ended in failure if we are to improve upon this technology.

References

DeBusk TA, Baird R, Haselow D, Goffinet T (2005) Evaluation of a floating wetland for improving water quality in an urban lake. In: Proceedings of 8th biennial conference (2005) on stormwater research and watershed management. Southeast Florida Management District, Brooksville, Florida, pp 175–184

Galli U, Schüepp H, Brunold C (1994) Heavy metal binding by mycorrhizal fungi. Physiol Plant 92:364–368

Göhre V, Paszkowski U (2006) Contribution of the arbuscular mycorrhizal symbiosis to heavy metal phytoremediation. Planta 223:1115–1122

Headley TR, Tanner CC (2006) Application of floating wetlands for enhanced stormwater treatment: a review. Auckland Regional Council Technical Publication No. 324, Auckland Regional Council, Auckland, pp 1–93

Hubbard RK (2010) Floating vegetated mats for improving surface water quality. In: Shah V (ed) Emerging environmental technologies, vol 2. Springer, Dordrecht, pp 211–244

John CM, Sylas VP, Paul J, Unni KS (2009) Floating islands in a tropical wetland of peninsular India. Wetl Ecol Manag 17:641–653

Kadlec RH, Wallace SD (2009) Treatment wetlands, 2nd edn. CRC Press, Boca Raton, p 1016

Li X, Mander Ü, Ma Z, Jia Y (2009) Water quality problems and potential for wetlands as treatment systems in the Yangtze River Delta, China. Wetlands 29:1125–1132

Mitsch WJ, Gosselink JG (1986) Wetlands. Van Nostrand Reinhold, New York, p 539

Nahlik AM, Mitsch WJ (2006) Tropical treatment wetlands dominated by free-floating macrophytes for water quality improvement in Costa Rica. Ecol Eng 28:246–257

Stewart FM, Mulholland T, Cunningham AB, Kania BG, Osterlund MT (2008) Floating islands as an alternative to constructed wetlands for treatment of excess nutrients from agricultural and municipal wastes—results of laboratory-scale tests. Land Contam Reclam 16:25–33

United Marine International (2010) http://www.trashskimmer.com/anacostia2.htm. Accessed 21 June 2010

Vymazal J, Kröpfelová L (2008) Wastewater treatment in constructed wetlands with horizontal sub-surface flow. Springer, New York, p 566

Index

A

Aapa peatlands, 98
Abiotic, 11, 29, 68, 214
Acer, 49, 50
Acidification, 211, 212
ACOE, 144, 147, 173–182, 184, 191
Acorus calamus, 242
Adaptive, 3, 5, 21–23, 27–29, 31, 33,
    211–213, 218
Administrative Authority, 196, 197
Aerenchyma, 239
Aerobic, 124, 238
Aerosol, 135
Aesthetic, 3, 5, 6, 12, 20, 105, 115, 116, 208,
    209, 214, 217, 223, 227, 228, 230, 233,
    237–239, 241
African-Eurasian Migratory Waterbird
    Agreement, 195
Agriculture, 35, 27, 38, 54, 61, 77, 105, 107,
    121, 189, 201, 207, 217
Agroecology, 131
Albedo, 135
Alder, 49, 55, 84
Algae, 135, 138, 228, 240
Alisma, 76
Allochthonous, 72
Alluvial, 28, 31, 33, 53, 80, 206
Alnus, 49, 50, 84
American Association of Landscape
    Architects, 224
Ammonia, 123, 237, 247, 249
Ammonium, 238
Amphibian, 107, 116, 125
Anacostia River, 23, 237–239, 242, 243, 248
Anaerobic, 124, 238
Animal, 8, 10, 38, 57, 59, 67, 69, 84, 97, 116,
    120, 124, 125, 178, 193, 214
Anoxic, 60, 123
Anthropogenic, 36, 38, 48, 53, 69, 82, 83, 132

Antietam Formation, 43
Aquatic, 97, 107, 108, 122, 125, 144, 149,
    175, 177–180, 237–241
Archaeological, 116, 205, 206
Architecture, 171, 223, 224, 226, 233
Archive, 53, 60, 67, 68, 74
Arctic Seed Vault, 134
Ardea alba, 208
Arkansas, 103, 194
Ash, 49, 80
ASLA, 224
Assessment, 74, 97, 101, 103, 105, 108, 109,
    137–139, 142–144, 148, 149, 166, 167,
    175, 180, 198, 201, 217
Astacus astacus, 211
Atlantic white cedar, 83
Attention Restoration Theory, 158
Attitude, 3, 6, 7, 12, 18, 200
Attleboro Mall, 177, 178
Autochthonous, 72, 79, 80
Avoidance, 95, 171, 174–177, 179, 182, 184,
    185
Avulsion, 32, 33

B

Bacteria, 123, 238, 240
Baffles, 225
Barkarra, 100
Base flow, 32, 44
Bedload, 32
Being effective, 157, 159, 166
Benchmark, 9–11, 60, 68, 83
Benefit, 6, 12, 13, 17, 18, 20, 95, 108, 109,
    116, 117, 119, 120, 125, 126, 165, 166,
    193, 199, 205, 207, 209, 212, 213, 215,
    218, 223–225, 231, 234, 240–242
Best Management Practices, 18, 118, 119, 224,
    229, 230
Betula, 49, 50